East End to South West

A life story

Ron Sambell

First published by Shakspeare Editorial, November 2022

ISBNs pbk 978-1-7397590-2-5
ebk 978-1-7397590-7-0

Dedication

To Paul who shanghaied my heart.

To my children Patricia, Stephen, Richard, Philip, Lindsay, Robert and their families, for whom I have enduring love and great pride.

Contents

Preamble

This is the life story of an 'anybody' with some family history thrown in, so don't expect any earth-shattering revelations. I have attempted to tell my tale through my eyes, as I age from being eleven. Why eleven? Don't fret, it soon becomes obvious.

My vocabulary as a child was simple and colloquial Cockney. To strengthen the narrative I have made numerous asides in my ninety-three-year-old voice – these interruptions are in italics and passing over them will not detract from my story. There is also some family history, background information and a better understanding of my psyche, also in italics.

My original title for this book was 'The Holes In My Boots'. It was meant to convey a sense of transit and a figurative sense of my life experiences. I have kept it alive by chapter headings and text.

Primary sources on my ancestors are indicated by an *. These include birth, marriage and death certificates; BMDs; census returns; parish records; invitations to births, marriages and deaths; and photos with information in the image or on the reverse side.

I've included photographs, genealogical lines and family trees.

I've taken the liberty of adding a few poems, free-writes and a challenge.

It may be of interest to start by chronicling some of the events that occurred in 1928, my birth year.

National Headlines	6-7 January	Fourteen drown as Thames floods London
	7 May	Representation of the People's Act, lowering the voting age for women from thirty to twenty-one
	3 July	Logie Baird demonstrates colour television
	20 December	First Harry Ramsden fish and chip shop opens
International Headlines	10 May	First scheduled TV programmes by General Electric
	15 May	First appearance of Mickey and Minnie Mouse in Plane Crazy, a Walt Disney film
	17-18 Jun	Amelia Earhart, first woman to fly solo across the Atlantic

Chapter 1

Early Days and the Family History Bit

With the clunk of closing doors and the waving of his green flag the guard gives the go-ahead. Belching smoke and with a screech on the steam whistle the train judders and pulls away, carrying hundreds of us kids and oldies from London to … I don't know where.

I'm on the quarter-to-three from Paddington. Heads poke out and arms wave from carriage windows. Mum is getting lost amongst the crowd of mums and dads we are leaving behind.

'Shut the window Ronnie,' Gran says in a snappy way, 'I'm getting covered in soot.'

I do as she says and sit next to her, my gas mask box digging into my back. The yellow label hanging round my neck tells anyone who needs to know that my name is RONALD SAMBELL, d.o.b. 11 Aug 1928, BRITISH group B. With me are my eight-year-old brother Alfie and our Grandma, she's our Mum's Mum. Gran, who is taking up two seat spaces, pulls a long hatpin out of her black hat and as she takes it off her greyish hair falls to her shoulders; the five kids sitting facing us stare goggle-eyed as she opens her big shiny bag and takes out a cloth to wipe her sweaty face. Looking up, I ask her where we're going.

Her shoulders shrug as she says grumpily, 'God alone knows son, we won't 'til we get there.'

Tears come to my eyes. I'm muddled. What's happening? Mum was crying as we all left our home in Old Ford to get the Number 8 bus to Paddington. Mrs Segal, the hairdresser next door, and Ruby, one of Mum's servers in our restaurant, were there to see us off and they were crying as well. Ruby had shouted out to us to be good boys for Grandma as the bus started. Why do I feel as if I have lost something? It's not like when my weekly sixpence dropped down the drain or when I have to give one of my toys to Alfie. Why the suitcases? No, this is different, something's going on.

It is 1 September 1939 and Operation Pied Piper, the mass evacuation of children, pregnant women and elderly people from major cities is underway; war with Germany will be declared two days later. This forced but necessary break-up of over one and a half million families will have lasting effects on most of us. Some will never be physically reunited, and many won't reconnect socially or psychologically with their families. Julie Summers writes in depth of these problems in When the Children Come Home.

Old Ford, in the East End, is named in thirteenth-century records; it was an important crossing of the River Lea on the major road from London into Essex, until Bow Bridge was built.

There's no corridor, so no wandering and no lav. With my head on Gran's chest some of my schoolmates come to mind. Bert, Lenny, Maisie, Sid. I've known them forever and they are all on the train somewhere.

But a chapter in my life was closing. I will never see any of them again.

We were all in the same class at Atley Road Primary. I think back to when we were titches and the five- to six-year-olds were made to rest on cots in the afternoons with the curtains drawn. They don't close proper and I remember lying there watching the dust turn into the sunbeams, itching to be up and out. Other things that happened regular were the nitty nurse combing hair looking for fleas and boils on the back of the neck; mine never came to much but some older boys had big ones which were dealt with by squeezing the pus out. It must have hurt like billy-oh. It makes me think of how much I hate taking Syrup of Figs but how great it is to swallow Galloway's Cough Mixture when my throat is sore.

The infant school finished at half past three and Ruby would be there waiting for me and most times some of my mates would string along. One or more of them would follow us into the restaurant to get a doorstep and dripping. Mum never turned them away.

Later in life I will understand that this need for something to eat was due to real hunger, the sort poverty causes. I will also come to know how much more privileged I was.

After school I'd play with my mates; swapping fag cards, playing glarneys in the gutter, whipping tops, hopscotch, skating, being bad when we knock-down-ginger. Maisie knocked spots off us when it came to skipping. On a wet day Mum would let me take my mates down to the playroom to play with my collection of lead soldiers, cowboys and Indians. Tiddlywinks, snakes and ladders and snap were also favourites. Mum always gave us Chelsea buns and squash.

On Saturday mornings, as infants, we went to the penny pictures at the Wesleyan church hall. We watched silent black-and-white flicks of Mickey Mouse, Laurel and Hardy, Harold Lloyd, The Three Stooges, Korky the Cat, while some old lady played along on the piano. Later, the Geisha in Parnell Road put on Saturday morning talkies for eight- to twelve-year-olds. The noise is very loud, with kids running around and jumping up and down. You never want to sit below the front-row balcony seats, the stuff those up there throw down. My favourites are Flash Gordon the spaceman, Tom Mix the cowboy and Popeye the Sailor Man, but I don't much like Crash Corrigan.

This Saturday morning treat is for taking the tablecloths from the restaurant to the bag-wash every Friday after school. I'd use a pushchair to carry the washing, with Alfie sitting on top. It was this job that resulted in a split fingernail on my right hand after the pushchair tipped back and caught my fingers against the pavement. The nail was torn off and my hand got infected.

No antibiotics in those days. Although Fleming developed penicillin in 1928 it was not generally available to GPs until 1945. It was touch and go whether my hand would have to be amputated. It had one good outcome though, teachers had to stop trying to make me use my right hand to write with. I still have the split fingernail.

On Sunday, outings in Dad's Austin 12, Epping Forest was a favourite, or a weekend in the caravan at St Osyth Beach. We listen to *Dr Fu Manchu* and the *Ovaltineys* on a battery wireless on the way home. A must during the week is *Dick Barton, Special Agent*. Over the last three years, summer holidays were spent in our caravan and my cousin, Baby Mary, and her Mum, Auntie Mary, my Mum's sister, would come as well. Baby Mary's Dad, Uncle Tom, has a penny-in-the-slot arcade on Clacton pier.

Talking of piers reminds me of the outing Alf and I made with Mum by paddle steamer from the Embankment to Southend pier. The bloke telling us about things as we passed them finished up by claiming it to be the longest amusement pier in the world.

Our caravan is in a field behind the sand dunes where us kids spent a lot of time catching grasshoppers and making hairy grass ears run up our arms. Dad and I always got up early for a walk along the beach to a stall for a mug of hot chocolate because it could be pretty chilly first thing.

The Family called Cousin Mary 'Baby Mary' to avoid confusion with her Mum; the cognomen stuck, even after her death at ninety-two. My Mum and Dad were called Nan and Nib by the family – Mum called him Nibby.

St Osyth, originally Chich, is on the north east Essex coast, about five miles from Clacton-on-Sea, with Jaywick Sands in between. It is recorded as the driest area in the UK. Our campsite has been developed into St Osyth Beach Holiday Park.

St Osgyth or Ositha, daughter of Frithewald King of the East Saxons, lived in the seventh century. She was beheaded by invading Danes for refusing to renounce her Christian faith.

When other aunts and uncles joined us at St Osyth, Alf, Baby Mary and I would sleep in our big bell tent. Once, when I jumped into bed, something shot up my leg. I must have hollered because Mum rushed in. She told me to stand still and yanked down my jimjams. Out fell a monster creepy-crawly. For a while after that fright I'd pull back my bedding and check, wherever I happened to be. We also holidayed at places where the beach is all sand and I recall the fun we had burying Dad up to his neck. He made it easy by lying down on his back. There were also donkey rides and Punch and Judy. What sticks in my mind is Punch shouting out, 'That's the way to do it', whenever he whacked all the others over the head with his stick, even his wife and the baby. Although the kids booed when he did this he was always warned with a 'Look behind you', when the bobby , the clown and the geezer carrying the gallows crept up behind him.

By the time I was nine or so I was allowed to go to the local library on my own. The first book I ever borrowed was *The Cloister and the Hearth* by Charles Reade. No kidding. Why I chose it I shall never know, talk about heavy going. But I did read it through. Reading is becoming a habit; I have read a cartload since then.

A Nod To A Novel

My love for literature is off the wall
The print on a page is a lasting passion
A single sentence can whisk me away
A poem well-phrased in a similar fashion
When the mobile library is in my locale
I'm there to the second to make my selection
Thriller, travel, sci-fi or textbook
My bookshelves abound with fact and fiction
The choice is to die for my hunger demanding
Classics crammed in to quench my addiction
Austen, Cooper, Gaskell and Conrad
I'll even read rubbish 'til my eyes go blurry
My brain gets befuddled but I still crave for more
This love for a story ... well it's really quite scary.

My favourite stories are about: Blackshirt, a lah-di-dah crook and adventurer; Biggles, he's a flying ace; Just William with Violet Elizabeth, they're just a great laugh; then there's Bulldog Drummond and loads of others.

I recall a visit to the library which got me into hot water with Mum. I have to tell her what I'm doing if I'm not coming straight home from school. This time though I hadn't said. I was well over an hour late and got back to see a small crowd and a bobby standing outside our door. Then some man spots me and shouts, 'There 'e is. Blimey kid you're forrit.' I knew then I really was in for it.

Mum ran up to me in a right tizzy and yelled, 'You bad boy where have you been?' I got a good shaking and a slap on the leg before being grabbed hold of as she started to cry.

This was the first of two occasions in my long life where my thoughtlessness caused a lot of angst for a cherished one. We will come to the other one much later.

Grandma started our restaurant in 399–401 Old Ford Road but Mum owns it now. It is on the slope of the bridge over the North London Railway. Ruby, Dot and Iris are her waitresses. They are great fun and always sing the latest songs for the customers. Flo and Maggie work in the kitchen with Mum, they're older and don't have much to say. Most customers come from two very big factories on the other side of the railway – one makes wallpaper and the other one matches – and from Old Ford Station, which is across the road, as well as Mum's regulars and passers-by.

It is called the Railway Dining Rooms and is two shops joined up to make one. One front room has fixed seats and tables for ordinary workers; the other has a separate entrance, padded chairs and tables with cloths laid for four and is for the more well-off people who run the factories and the station. The menu is the same for both sides, it's just that the snooties usually have afters and pay a bit more.

After one Christmas party Mum did for the factory bosses we went round the tables, drinking leftovers from the glasses. Dad went doolally.

We live in the rooms upstairs. There are three big bedrooms and an even bigger parlour. In my baby years there was no bathroom and the lav was in the back yard. Among my earliest memories are madly itching chilblains, pee pots under the beds, a tin bath in front of the fire on Friday nights and sitting on Gran's lap in a corner of the big kitchen during busy times. Gran's cure for itching chilblains is to put my toes in the pee pot after I've done a widdle. It works like magic, but afterwards my feet are washed and it's not long before the itching starts up again.

All our family meals are eaten in the main kitchen. Alfie and I sit at a separate bench-top and are not allowed to talk. We often whisper and have giggling fits, which can lead to Dad marching us out with a, 'Go upstairs'. I don't know why he is so strict about all this. He even shut me in the chicken run once.

This isolation and silence at mealtimes was the norm. I wasn't aware that the opportunity to bond as a family was being lost. It could account for my difficulty in maintaining everyday chat, apart from now being very deaf.

The back garden is on a level with the railway so there are two rooms under the restaurant – one is the playroom and the other a stockroom. And there are two cellars: one is used as a scullery –there's a coal-fired copper in one corner, a deep sink with a scrubbing board and a big mangle with wooden rollers and the machine for scraping the skin off spuds; the other is the coal-hole, full of long-legged spiders. The spiders didn't bother me at all when Alf and I played at being coal miners, but then Mum and Dad took me to see an adventure film at the Geisha. There is a bit where a boy runs into a cave and gets stuck to a big web and monster spiders start creeping towards him. It scared me something rotten and I stopped playing in the coal-hole after that. Though what really put me off going down to the cellars without having all the lights on was being taken to see *Frankenstein* for my birthday. I took the stairs two at a time coming up after that. They made up for it by taking me to see *Show Boat*, the song 'Ol' Man River' keeps going round in my head.

Spiders have given me the shivers ever since. Not getting scolded for being covered in coal dust was a bonus I guess. God knows why they took me to see Frankenstein, a 1931, pre-code, black-and-white horror film. Believe me when I say it really put the frighteners on me.

Show Boat was my first face to face with two things that were to have a strong influence on my values. First, slavery and its implications for equality; Dad bought the sheet-music of 'Ol' Man River', as performed by Paul Robeson, who was also an opera singer. Jerome Kern's lyrics are explicit regarding slavery. Second, the operatic quality of singing in popular entertainment; besides Show Boat, I like The Chocolate Soldier (Guardsman), Porgy and Bess, West Side Story and The Three Tenors. A third was Frankenstein and the consequences of obsession, of which Hitler's Third Reich and the concept of a master race is a horrendous example.

Later, I added a fourth cinematic influence on me; in 1940, Walt Disney had produced the film Fantasia, an anthology of classical music set to animation. Of the seven pieces portrayed, it was Beethoven's Pastoral Symphony that struck a chord with me and triggered an abiding interest in the genre.

By the time I was about seven Dad had turned the stockroom into a bathroom, but the lav is still outside. I still hate using it. It empties in the usual way by pulling the chain, and there is a holder for the scratchy Izal toilet paper.

'So what?' you ask. But have you ever had to go out in the winter and plonk your tender bum down on a clammy, fixed plank seat? Well, have you? I think not.

Our parlour has a big open fireplace, good for toasting the crumpets and roasting the chestnuts Mum buys from the muffin man. He comes round every Sunday with a big wooden tray balanced on his head and ringing a bell as he shouts, 'Muffins for sale, penny a bag chestnuts'. Other favourites on Sundays are cockles in vinegar and winkles. Once boiled, the 'foot' of a winkle is black and hard, after pulling the insides out with a pin and eating them, we'd put the blackened feet on our faces, do this enough times and it looks as if you have some deadly disease.

Because of its' size, the parlour is often used for a Saturday evening knees-up. These usually include Auntie Mary, Uncle Tom and Baby Mary (two years older than me and a right looney). Others who are most likely to be with us these Saturday evenings are Grandma's sister Hannah and her husband Fred Simkins, and one or more of their five grown-up boys; also her other sister Louisa Dodsworth, she isn't married and lives with Aunt Hannah and Uncle Fred. Until I was about six my Grandpa, Alfred Sambell, would come over. He was not very tall but nearly as wide, with a bristly moustache above a smiley mouth and he always threw me up in the air with a twist when he arrived.

Grandma's husband was James William Low. He was losing his sight and had a fear of going completely blind. He shot himself in the head, 'being of unsound mind', on 8 June 1911. They had three children – Robert, Mary and my mother Hannah. Grandma Low (I have never known her as Grandma Long, although her death certificate shows her as married to Ernest Long) died on 1 May 1953*.*

Grandpa Sambell came from Morice Town, Devonport, Plymouth. He was a plate-laying riveter in HM's shipyards, until he moved with his family to West Ham. He was very strong and once, instead of standing watching as two men tried to get Dad's piano upstairs, he took over. 'Put it on my back,' he ordered. Then he climbed straight upstairs and plonked it down in the parlour. I can still hear him saying whenever anything was out of the ordinary, 'Well, I'll go to the foot of our stairs'.

Dad is about as tall as Grandpa Sambell but I don't think he's as strong, and he doesn't have a 'tache. But he does have the same bright blue eyes, the jutting jaw and short brown hair. He is great on the piano and his party piece is 'The Laughing Policeman'. He laughs like crazy, the veins stand out on his reddened neck and tears run down his cheeks. This makes the rest of us split our sides.

He has to do it every time there is a get-together; he pretends not to want to but he always sings it.

The Saturday before we were evacuated was to be the last time a family gathering of this sort would occur.

Mum's brother, Uncle Robert, and his wife Aunt Alice, run a cafe on the other side of the River Lea. They're not far away but I hardly see them. They have three kids who call in sometimes, my cousins Gerald, Jimmy and Roy.

As personalities, Dad's side of the family, apart from Grandpa Sambell, are a mystery to me.

*My efforts to research Dad's side of the family have been helped by: some of the myriad photos Mum and Baby Mary's daughter Helen Doherty passed on to me; various invitations to births, deaths and marriages; and the England and Wales censuses for Grandpa Sambell's household. He was born in 1873 and died on 27 February 1934.**

Grandma Johannah Sambell, née Shakespeare, was born in 1875 and died when I was a toddler, on 19 April 1930. They had three daughters and two sons: Lucy Mary, Edith, Amy Ethel, Alfred Arthur (my Dad), and Charles David Shakespeare.**

*Amy married William Francis King, a chef; she died aged twenty-eight on 5 March 1934. Her address was 24 Shrewsbury Road, Forest Gate, Essex. She is buried at Manor Park Cemetery.**

Dad is doing alright. At the age of fourteen his very first job was at a grocer's shop. Now he works for Buck and Hickman. They make small tools in a factory down in Whitechapel Road. From starting there as an 'ironmonger's assistant'* Dad is now something called a Floor Supervisor.

For some reason I shall never fathom, my father was a Freemason and was jointly instrumental in creating the Resergum Lodge at Liverpool Street Station. He retired from Buck and Hickman in 1967 as the Sales Director. By and large he wasn't a joyful man and his skill on the piano was a bit out of character; although I do remember him coming home* *one time after his work's Christmas party dragging a large plucked turkey behind him and telling it to 'heel'.*

Dad is good at mending around the house and making things for Mum. From sawn planks to shining tables my Dad could work wonders. Before I started going to school, I remember Dad had made me a boat. Ruby would take me to Victoria Park and we'd sail it on the lake. It had something Dad called a keel on the bottom to stop it tipping over in the wind. It's funny how things look when you are small; Ruby would let me go on the swings which had seats with bars so that toddlers didn't fall off. Those for the bigger kids were just plain seats on which they would stand, they then worked them to and fro until very high and I would cry with fright for them. Now I do the same thing. For my eighth birthday he made me a posh wooden scooter, I was in heaven. The wheels were two iron ball races; what a great clattering sound they made going over the cobbles. It meant I could join in with the gang, who all had home-made scooters or trucks of their own. I'd soon worn a hole in my boot, the left one of course.

I feel very sleepy as Gran shifts my bonce and I wonder whether Mum will look after my sets of fag cards, my bag of glarnies and my collection of *Boy's Own*, *The Wizard* and *The Dandy* comics while I'm away. My most favourite stories are about Wilson in *The Wizard*. He is a super-sportsman and adventurer and wins or sorts out everything.

'I reckon she's already chucked them out,' Gran says.

I see she's grinning so I don't think she means it.

1886 Mary Low, born Horsewill in 1839. My great-great-grandmother

1892 ca Mary Dodsworth. My great-grandmother. She married James William Low on the 25th of December 1894. A Christmas day bride

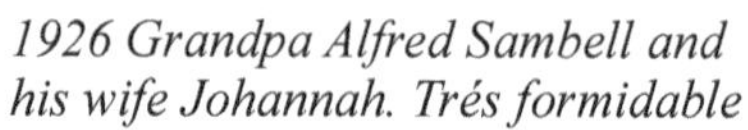

1926 Grandpa Alfred Sambell and his wife Johannah. Trés formidable

1880 ca Mary Horsewill, on the left, and her family

1918 Alfred Arthur Sambell, age 16, in apron on the right. My dad to be

1921 Grandma Low's Christmas card to Grandpa Sambell and his family. (It tells me Mum and Dad had a long courtship)

1923 Grandma Johannah Sambell born Shakespeare. 19-year-old Mum on the left, Dad's sister Amy on the right

1920ca Old Ford Road. Nearest shop on left is hairdresser, Mrs Segal, then our double-fronted restaurant and then a sweet shop. Old Ford Station on the right

1925 My Mum to be, Hannah Horsewill Low, on her 21st birthday

1923 My Dad, Alfred Arthur Sambell. His 21st birthday

1924 Grandma Mary Low with her two daughters. My Mum Hannah on the left with Dad, her steady; Aunt Mary on the right with fiancé Uncle Tom Everett

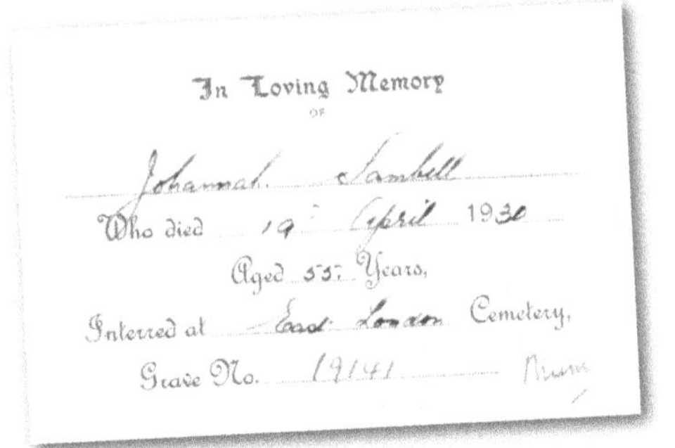

In Loving Memory
OF
Johannah Sambell
Who died 19 April 1930
Aged 53 Years,
Interred at East London Cemetery,
Grave No. 19141 Mum

1930 An example of a notice that allowed me to build my family history. My Dad has written Mum in the corner

1927 June 5th. The wedding day of my Mum and Dad (Nan and Nib)

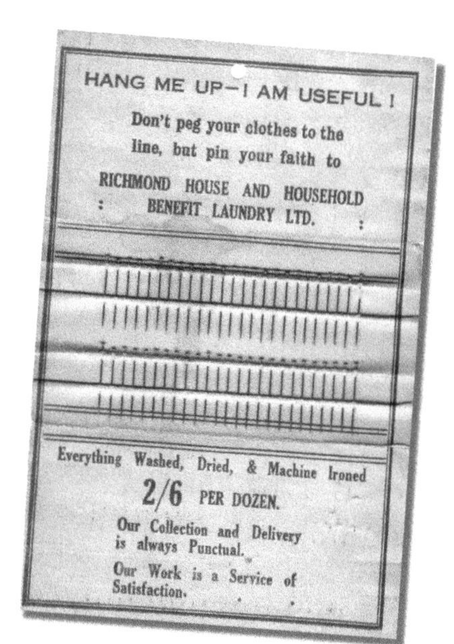

1930s A typical advert for services plus a free card of pins

1930 Baby Mary and me burying Dad in the sand. It became an annual ritual

1933 My 5th birthday with Alf

1933 Alf and me, donkey rides on Margate Sands. Getting nowhere fast

1936 The three chefs-cum-bottle washers: Brother Alf, Baby Mary Everett and me at our St Osyth family campsite

1950 Nan, my Mum, on left outside her reastaurant with Ruby and a railway official. Empire billboard shows Lucille Ball in'The Affairs of Sally' released in 1950 in the UK

1936 Mum, Alf and Me on the pleasure cruise from London to Southend

Chapter 2

Townie Cum Bumpkin

The train jerks as it slows down and we all start shuffling round as we pull in at a place called Reading and dozens of us with white or yellow labels are told to get off. Alf's hand is tight in mine as I drag a suitcase with the other. What a crowd; pushing and shoving, waiting for orders from a red-faced bloke in a bowler. He yells at us to stop talking and listen.

'Those with white labels, get yourselves in a line next to me.'

There's a lot of mix-up before it's sorted out. Then it's the yellow labels' turn, we have to form a line in front of him. Having got this done he pushes off. Alf is tugging on my sleeve, he sounds a bit desperate when he says he's bursting. I think it's best if he does one so I tell him no one is going to notice if he just goes and pees against the wall. Of course, eagle-eyed Gran catches him and I cop it, I bet she's in the same boat though 'cos I know I am.

After a bit of a wait bossy-boots comes back wiping his 'tache with a bright-blue snot-rag. He tells us those with a label marked Group B are to go to the first two double-decker buses waiting outside the station. A look at my label tells me it's us and we get on the first one. Gran is finding a seat near the door as Alfie and I make a beeline for the back seat.

What follows is five years of discovery and rapid development.

Spencers Wood is where we get off the bus. I feel very tired and homesick now. I think we have been able to keep together so far because Gran is with us. Even so, I find myself dumped on a farmer while Gran and Alfie have ended up at a house on the main road.

Am I in a new world? The farmer has got to be a giant. As he opens the door wider I'm staring at the big brass buckle of the belt holding up his baggy trousers. I look up at his face, his coal-coloured eyes stare down at me.

'And you're … ?' he asks in a funny sort of voice.

I say my name and add I'm the evacuee from Old Ford. He frowns and says he is Paul Coran,

'Where'd you say you're from?'

Before I can answer he takes my suitcase in a large fist of a hand and tells me to come on in. I must look as frightened as I feel because his bushy eyebrows lift and he laughs. I'm really tired now and I start to blubber. Through the tears I see someone is now standing next to him, her head cocked over, touching his shoulder. Giving me a smile she says she's Mrs Coran and I can call her Marie. As I follow her through she gasps and turns to Mr Coran, puts her hand to her mouth and with a worried look asks him, 'What's this the cat's brought in?' Mr Coran says if I turned sideways I'd disappear. Then she smiles again, touching her frizzy fair hair, as she tells him that with my curls I could give Shirley Temple a run for her money. I blush and fight hard to stop the tears, only to make my nose run. She leads the way into a big kitchen where the warmth and cooking smells make me think of Mum. They are about to have tea so Marie tells me to take my coat off and where to sit.

As I eat they tell me they have three boys. Dan the oldest is already in the Navy and Colin has just been called up. The youngest is Gordon, he is here with us at the table. He is nearly fifteen, a lot taller than me and beefy, with floppy black hair all over his head. He'd left school when it broke up for the summer holidays.

The school leaving age was fourteen and not increased to fifteen until 1947.

Gordon swallows a mouthful and tells me they've only been here a couple of years themselves. He helps himself to some more rabbit pie as he adds that they come from Guernsey. Where on earth is Guernsey I'm wondering as Marie chips in.

'Yes, and we brought with us a herd of twenty-five Guernsey cows, a bull and Moses.'

A couple of days pass and I'm still trying to take it all in when Mr Coran leans across the kitchen table. He says I can call him Paul then he asks me, all serious-looking, if I'm up to milking tomorrow morning. Crikey, I don't know one end of a cow from the other but before I can answer Gordon puts in his tuppence-worth to say not to forget the mucking out afterwards. I catch Paul giving Marie a wink and they all have a good laugh. Marie says to pay no heed, I'll soon pick it up; in the meantime, I shall be giving Gordon a hand. Paul shakes his head as if I'm getting it easy and tells me I need to get a good night's sleep because I shall be called at 6.30.

Gordon makes no bones about yanking me out of bed at the crack, well, 7.30-ish. Out in the farmyard my eyes squint against the bright daylight. They've already done the milking and Gordon says my first job is to harness Moses to the cart. Yep, Moses is an enormous horse, a Shire he tells me. I jump at his offer to help me because all I can see is a jumble of straps and gleaming buckles hanging in the stable. He fetches a crate for me to stand on and takes me through the harnessing of Moses. He loads churns of milk on the cart and says I'm to take them down to the lane and wait for the lorry, where the driver will exchange them for empty ones.

'Don't forget to bring back the docket,' he says, wagging his finger at me.

What the heck's a docket? Sitting up behind Moses, I haven't a clue about driving a horse but Moses has it all down to a T. Tom Mix comes to mind and I decide to be a stagecoach driver fighting off Red Indians. I slap Moses' backside with the reins and whoop, but it makes no difference to his slow plod. The lorry turns up and, before he leaves, the driver gives me a piece of paper; now I know what a docket is. Back in the farmyard Gordon takes me through the job of feeding the pigs, hens, ducks and geese and then, not before time, we're going in for breakfast.

Gordon was my mentor. He was a gentle giant of a boy who took me under his wing and patiently introduced me to the hustle and bustle of a busy farm. Perhaps he saw me in the same light as the animals. There's no gainsaying the fact I didn't know a duck from a goose ... or a cow from a bull for that matter. Even as I write, the enfolding smell of the cows' warm bodies bedded down in rustly straw on a winter's night is right back with me.

As I eat my bacon and egg Paul sits back. He tells me the harvesting will be late this year because of the wet. He thinks I will come in real handy now the boys are gone. What's that all about? I'll soon find out, I suppose. Sure enough, the bundles of wheat, sheaves they say, are being chucked up on to the cart by Paul and Gordon and I'm putting them tidy to get most on. Back in Home Paddock we build the sheaves into large stacks ready for a visit from the owner of what they call a threshing machine. Four days of harvesting are behind us and now I watch this monster gobble up sheaves at one end and spurt out seeds and straw at the other.

It's 31 October 1939 and Gordon's fifteenth birthday. Marie puts on a party for him and people turn up to add to the fun. He receives lots of presents and I feel a bit lost. As the last ones leave Paul turns to me, 'You look a bit sorry for yourself, Ron.' He laughs, 'Don't worry you haven't been forgotten.'

I shuffle a bit but he's hit the nail on the head. His last words perk me up though. He pokes me in the ribs and points to the door, Marie and Gordon are standing there all po-faced. She calls for me to get a move on and shoves me towards the scullery. I look in and there, in all its glory, is a bike, decked out in ribbons and tinsel. Gordon asks me what I think and reckons its miles too small for him anyway. They all start laughing and I'm lost for words. I stutter out my thanks as I look at them in turn.

I can't ride a bike. Two days of getting on and falling off, feeling daft. Then suddenly I have my balance and there's no stopping me. This bicycle is going to be a life saver. It's a must, now petrol is rationed and there aren't many buses. I've been walking the two or so miles to school ever since I got here. I've not been looking forward to doing it in the winter.

Yes, I go to Lambs Lane School. It is on the edge of Spencers Wood going toward Basingstoke. Kids from three nearby villages come here as well. It has an office and two big schoolrooms in which four classes used to be held. Two out of the four teachers have been called up, so now there are only two classes: for five- to ten-year-olds and for eleven- to fourteen-year-olds. The younger ones are taught by Mrs Wyeth, a stern-looking lady, but very gentle with the littlies. She and her husband own the bakery in Three Mile Cross, the next village to us on the way into Reading. Mr Wellow, the headmaster, is a fat, bald, old man with a soft voice that he never raises, but golly is he good at dishing it out with a ruler. He takes us older kids, there are about twenty-five of us. The lessons are a doddle, I already know most of it: times tables up to twelve, pounds and ounces, pints and gallons, inches and feet, sums, dictation, reading out loud and spelling bees, especially Hangman.

But why up to twelve times you might wonder? Suffice to say there were twelve pennies in a shilling, there are twelve inches in a foot and three times twelve inches in a yard, thirty times twelve degrees in a circle, a dozen is twelve of anything, there are two twelve-hour periods in a day and an hour consists of five times twelve minutes. Seriously though, I would begin to forget my English, maths, history and geography, the core subjects.

I become mates with Harry. He has trouble with the lessons but he makes up for it by being into all sorts of larks. Scratching the straw-coloured hair falling over his squinty eyes he tells me Welly has this thing about pencils. If you ask to sharpen them for him you can spend a good half-hour in the playground. Mind you it has to be done with a sharp penknife and it's got to be spot on. Welly is most particular about the finish.

The big tree in the playground is loaded with shiny conkers. Being in the conker championship turns out to be a must and Harry knows a trick or two. Pulling me close he whispers in my ear, 'You have to save some from last year, make a hole right through first and then pickle them in vinegar, which makes them go hard.'

This is where I find I can dangle a conker on a piece of string without pulling away as one of the others tries to break mine with their strung conker. It really hurts if the other twerp misses and hits your knuckles, but pull away and you're called a pansy.

The lavs are in an outhouse away from the school building. There are lavvy seats but you have to do it into buckets. These are emptied on the compost heap in the school allotment by old Robinson the caretaker. I don't know who gets to eat the veg. Us I guess, in the school dinners.

Winter is setting in early this year. Yet another snowstorm and I'm struggling home pushing my bike. School had finished at three as the coal had run out for the schoolroom fires. The rutted ice on the road is now too dodgy to bike on. I should have listened to Gordon. Don't be daft he'd warned me, it's a lot worse than the tramlines in Reading. You'll be better off not having to push your bike. Well, I hadn't listened. I'd wanted to show I'm the bee's knees.

'That's a joke,' I grumble aloud as I struggle with the bike. 'I've got two skinned ones instead.'

Two miles later I'm back at the farm and head straight for the kitchen. The smashing smell of baking is in the air. Marie is knocking-up the Christmas puddings. She shoves hair off her face, puffing as she tells me to come and have a stir and a wish. I grin, my eyes roll upwards as I slide over and do as she says. As she turns and stretches for the cake tin I beat her to it and reach over to get it. She smiles her thanks at me as she opens it and tells me I'd best have some before bringing in the cows.

How It Was

In my Mother's kitchen
The cupboards were full
The dresser well-laden
Jars and cans
Sachets and bottles
Powders, crystals, syrups and cubes
Evocative aromas of spice and seasoning

In my Mother's kitchen
Two iron-clad ovens
The grate between them glowing brightly
Shrilling kettle
Bubbling pots
Flour and water, butter and currants
A tray for the scones greased and waiting

In my Mother's kitchen
The windows were steamy
A pushed back cowlick
Rosy faces
Lick-sticky fingers
Redolent smells of nutmeg and cinnamon
Christmas pudding a wish while stirring

In my Mother's kitchen
Our woes were sobbed out
Soft voice of compassion
Shsh now
Never you mind
Her love spilling over pervading the room
Fragrantly airborne like the herbs at the ceiling

The idea of something being 'fragrantly airborne' is lifted from a passage by Nigella Lawson.

It is snowing again as I leave the kitchen. Flick, flick, blink, blink, the fine flakes skim across my face. Oak Meadow is three fields away and as I trudge on, the daylight goes and the drifting snow covers everything, making landmarks difficult to see; the silence of it all bothers me. Luckily, the herd is waiting at the gate and soft mooing guides me towards them.

A jittery thought crosses my mind, 'What if I gits 'em lost on the way back?'

'Don't be so daft,' I grouse aloud. 'When did you ever lead 'em back? Mattie does that.'

The bell hanging from Mattie's neck starts clanking as I unlatch the gate. She gives it a good barge, dumping me in the snow, my ankle twists as I go down. Off she trots, heading for the farm, the rest of the herd follow the sound of the bell. The last one out is wonky-legged Doris; as she passes I grab hold of her and struggle back through the drifts and into Home Paddock, where a dim waving torchlight comes nearer, it's Paul and Gordon.

'So,' Gordon says with a grin as we meet. 'What kept you then?'

'And the cows?' I gasp, as my weight goes on my swollen foot.

Paul gives him a push as he says they're all back and tells me he's well pleased I'm safe and I've done a good job.

It was the day I think I matured a bit. I most certainly felt as if this was where I belonged.

Christmas 1939, Gran and Alfie are on their way over because Mum and Dad have come. They look knocked out. This is their first chance to see us, bringing with them presents and their saved-up sweet rations. As well as bits and pieces of clothing, I look at a box labelled Meccano and wonder what it's all about. They know nothing about farming and aren't saying much about what's happening back home. It is a gloomy time for all; the only interest being us boys. Still, Marie does us proud by putting on a big roast Sunday dinner. As they say their goodbyes and catch the bus back to Reading Station, Alf has a bit of a cry, but I'm not feeling left behind.

We won't see them again for five-and-a-half years, not counting a brief half-day by train in 1944. This long separation was partly due to petrol rationing. It came into force in 1940 and petrol was soon restricted to official travel. Petrol rationing for the general public was restored in 1945 and limited to 180 miles per month for engine capacity up to 13 horse power. Pool Petrol, as it was called, cost 18d/gallon (less than 1p per litre). Food scarcity also comes to mind. I didn't see, let alone taste, a banana or a pineapple for nigh on five years .

What I don't yet realise is that my allegiances are transferring from my parents and the back alleys of the East End to my newly found family and the countryside.

In July, following Mum and Dad's visit, school breaks up for the summer holidays and I work all hours to help keep things going. Gordon teases me by squeezing my arm and asking how my muscles are coming on and will I last the course. I feel my arm and try to bulge the muscle. I tell him it's coming on okay, it must be the extra rations. I've been hoeing between the cabbages all morning and I'm hoping to have the afternoon off. A shout and I look up. It's Marie. She says my eats are ready and there's a letter from London. The hoe is put down pretty sharpish as I turn to follow her in. She gives me cheese sandwiches, a bottle of Tizer and the letter.

Taking it all to the bench by the pond in the paddock I flop down. The sun is hot on my neck and the geese have stopped nagging. A breeze moves the withy branches dipping in the water. The tiny waves this makes flash as they catch the sun. I lie on my belly and watch the sticklebacks and water boatmen. The letter is from Mum and Dad. Things are looking bad, they both had something to do with what's being called 'The Miracle of Dunkirk'. Dad has asked Grandma to look out for a house to rent.

'Why do things have to change?' I ask myself.

Well, some things change, others don't. I shall still be working on the farm and Alfie and I will still be going to Lambs Lane School.

Gran has found a house called Casseldene in The Square just up the road. Along with Alfie we will be moving in as soon as the house is empty. Casseldene is big and there is a walled garden all the way round. There are two parlours with wide bay windows, a dining room, something called a study, and a kitchen, pantry and scullery. Upstairs are four bedrooms and a bathroom and above these are the attics. We settle in and get used to it being lit by gaslight. Since there is no electric the wireless has a battery; like the one in Dad's car it is big and heavy and one of my weekly jobs is to get it swapped for a fresh one at Doubles garage, up on the main road.

Grandma has sent for Baby Mary. Being two years older than me she has left school and Gran has fixed for her to work for the village butcher. She's a bit of a tomboy with her black curly hair and twitchy grin. It's not long before the three of us are in Grandma's black book.

'Come on you two let's play Commandos,' Baby Mary orders. 'It's your turn to lead, Alfie.'

'How about parachuting into Germany?' he says. 'We could use a sheet and jump off the porch.'

Alf runs upstairs to his bedroom. As I follow he climbs out of the landing window and on to the flat roof of the front-door porch, carrying the sheet off his bed. Before we know it he is holding the sheet by the four corners and jumping. Panic stations. Taking the stairs two at a time I yell to Gran that Alf's hurt himself. I get to him to see his forehead is swelling up like a balloon and he isn't moving. Gran takes one look at Alf, orders me to get a cold wet towel and Baby Mary to get the Doctor. Alf is coming round and sitting up as the doc arrives. It all turns out okay, but black book tick number one.

Ticks in the black book mount up: Alf falling out of a tree on the common and getting a hairline fracture of his arm; Baby Mary chopping back the laurel bushes willy-nilly; me going through the ice on the village pond; scrumping pears from next door's orchard. To cap it all, we've collected so much wastepaper, one of our war-efforts, that the loft in the church shed collapses under the weight. I get the blame again. Each and every time I get a double tick and a good telling off, it's not fair. Every time there's a tick, Gran takes something out of our weekly sweet rations. She's an angel after all though. On Guy Fawkes Day she hands them all back. On top of that a bigwig from Reading awards us thank you certificates for knitting comforters for the troops and the paper collecting.

It's only early afternoon but everyone is looking up towards London. There's a bright red glare in the sky, which is gradually covered by thick clouds of smoke. It reminds me of a night back home when I was eight. There was a bright glow in the sky and Dad lifted me up to the window to watch it. The next day we found out Crystal Palace had burned to the ground. I don't know how it could be because Dad's paper said it was all cast iron and glass.

The redness in the sky over London was on 7 September 1940, Black Saturday as it came to be called. Hitler had decided he couldn't cripple the Royal Air Force so set about attempting to destroy British morale with 300 German bombers and an umbrella of 600 Messerschmitt fighters, dropping 340 tons of bombs plus incendiaries on the Dockland area of the Thames. The furnace effect I saw that day was the burning of the Royal Docklands and Poplar, 448 civilians were killed and 1,600 wounded. It was the first of fifty-seven consecutive nights of systematic destruction of Inner London and the East End. The last and deadliest day came later, on 10 May 1941.

Mum has since told me Dad was very lucky to come out of the Blitz alive. As an Air Raid Precautions Warden he survived two very near misses when bombs hit the two factories, Old Ford station and the railway line. He wasn't so lucky when later on in the war a German V-1 pilotless plane carrying blast-type explosives hit Buck and Hickman's. He was caught in the blast and limped home minus one of his lace-up shoes, which had somehow been torn off, and with his head covered in bandages. These doodlebugs, as they were nicknamed, were flying bombs. Thousands were aimed in the direction of large targets, usually London or other strategic areas, they simply fell to earth and exploded when the fuel supply was exhausted and the guidance system cut out. They made a threatening rhythmic throb going over. If the sound stopped you knew one was about to dive.

It is 11 August 1941, I'm thirteen today and, lucky me, I get two birthday parties. Early in the afternoon, Paul calls a halt to work on the farm because Marie has put on a do. Before we tuck in she gives me a balaclava helmet, which she knitted herself. Paul gives me a belt with a big shiny brass buckle and Gordon hands me his saved-up sweet ration. He gives me a sly grin and tells me not to be stingy and to hand them round.

'Not on your life,' I huff and smirk. 'Give-and-take is what they say and that's what I'm doing.' All the same, I do hand them round.

As a special treat he's taking me into Reading to see *Pinocchio* at the Vaudeville in Broad Street. Back home, Gran gives me another party, to which some mates from school come. Various presents come my way: a book from Gran; some sweets from Baby Mary; a catapult from Alf; a tin of spam and a half-a-dozen eggs from Harry; and a pet rabbit from Rodney across the road. All-in-all I have a happy day.

Christmas has come and gone. It was a sad time because Baby Mary's Dad, Uncle Tom, was killed last April during the Blitz. He was rescuing patients from a hospital in St. Pancras when another bomb hit the ward he was in. Auntie Mary's restaurant in Poplar was hit the same night and she came to live with us. I shall never forget the day she arrived. Her face was white and, as she stepped through to the hall, there was a long shuddering sigh as Grandma took her in her arms and stroked her hair. Auntie was in a right state and we've all been trying hard not to upset her.

Auntie Mary has been with us six months or so and things are getting back to normal. It's a cold night with no moon and it's good to be snuggled up in bed. I say my usual prayer for the safety of Mum and Dad. Before I know it, an almighty explosion jars me awake and then nothing for a few seconds. My door flies open and Gran yells at me to get up and take cover under the stairs. All stays quiet and about an hour later we stagger back to bed. Later, the milkman says a parachute mine had exploded at the crossroads in Three Mile Cross.

We all troop off to have a butchers. The mine dropped in a field, so there's no direct damage, just a great big hole. But the blast did damage several nearby cottages. Auntie Mary reckons the blast threw her out of bed.

London was badly damaged by the Blitz, but the most bombed city was Hull, where 95% of the buildings were flattened or beyond repair.

Gran tells me there's a letter from Mum and Dad. From what she has told them they think we aren't learning very much at Lambs Lane, so they are looking out for a school in Reading. Oh ho, things really are about to change.

Chapter 3

A Proper School ... and Other Things

It's the second week of 1942 and Alfie and I are wearing second-hand school uniforms decked out in the school colours of gold and brown stripes. We're getting on the bus to Reading for our first day at the new school. Alf is grumbling about the change and asking me what's it going to be like and saying that he feels a bit daft in the uniform. I tell him to buck up, it's my first time too.

Standing at the front gates of Reading Collegiate School we look at a big three-storey building. A sign tells us to use the back entrance to get to the office. Further back is a large gravel-surfaced playground. As we enter the office a tall, starchy man wearing a black gown turns towards us. He sniffs up his thin beaky nose and tells us he is the Headmaster, Dr Harmshaw, before asking for our names. I tell him and he frowns as he says Dad informed him we'd been going to a village school. He thinks this will have done us no favours and we will have to take tests to find which classes to put us in. Dad's feeling and what the Head has said are bang on, we are both placed a year lower than most boys of our ages. I'm starting in Year 4a.

Collegiate School is for boys going on to some sort of commercial job so there is no science. As ever, there is a shortage of teachers due to call-up. The fourteen-year-olds to the leavers are taught by Harmy for English and literature, Steero Steer for maths and sports, Foxey Fox for geography and history and gentle, white-haired Mr Heath for religion. Mrs Harmy does economics and business, plus shorthand and typing if wanted. Sports are compulsory on every Wednesday afternoon. For this reason there is school on Saturday mornings, now that does really take some getting used to.

As we settle in, Gran begins to worry about the amount of travelling we have to do. After the two occasions the bus doesn't turn up and we walk the four miles home, she finally writes to Dad telling him it doesn't make sense and he should think about getting us a house in Reading. He finds one in East Reading, at 12 Hamilton Road, a terrace house opposite the cemetery and close to Palmer Park – the Palmer of Huntley and Palmer biscuits, whose factory is in Kings Road. I apply to work there in the holidays. I shall get a few shillings for wages but the real benefit is the large bag of broken biscuits every Friday.

I can't for the life of me remember any proper farewells to the Corans. They were such an important episode in my early years yet, apart from a couple of visits by bike soon after the move to Reading, they were lost to me totally. An oversight I'm still ashamed of and greatly regret.

It is the first weekend following our move to Reading. Alf and I go exploring in Palmer Park. He heads for the playground but there is nothing there to play on, every bit of metal has been taken away for the war effort. Even the railings surrounding the park have been salvaged, only those round the local public library are still in place. We decide to have a go at climbing over them. It's pretty easy and just as Alf drops to the grass by me two girls turn up. Without saying a word one of them starts to climb over. As she turns to let herself down one of her a knicker-legs gets caught on a spike. Being the gentlemen we are we roll around laughing our heads off. She asks her mate to give her a hand and, raising her voice, she says.

'Those two have got it coming Aud.'

Aud gets over and helps her down.

'Okay, but it is funny Pip.'

They want to know who we are and, by way of a joke, I tell them Alf's name is Zebadiah and mine is Rozzie. They say they are at Alfred Sutton School for boys and girls, with teaching up to secondary level. When I tell them we are at the Collegiate they blow raspberries and pinch their noses. What's this all about I wonder?

The rest of my life had just been determined. Meet the future Mrs Pauline Patricia (Paul, Pip, Paula) Sambell, born Wilson. Paul's impact on my life cannot be too greatly emphasised and she is woven lovingly into all that has happened to me since.

My jokey name didn't stick but, to this day, Alf has been Zeb to my family.

Head Prefect and Cricket Captain Smailes calls out my name. He is head and shoulders above most of us so there's no problem spotting him. I go towards him wondering why he wants me.

'What's your cricket like Sambell?'

'I've never played it,' I murmur.

'Here, catch this,' he orders, throwing a cricket ball to me.

My hands go up and I duck my head to one side but, glory be, I catch it. That stung, I tell him.

'You'll do,' he grins. 'Turn up at the nets on Wednesday.'

It seems I have the knack and get picked to play against Presentation College. Against Reading School I hit three sixes. Harmy stops me on the way into class.

'Well done Sambell,' he says. 'It appears you saved the day. Keep it up, there's a good chap.'

As I lark about in the playground my House Captain, Rogers, stops me. He asks what I've been up to as he grabs hold of my ears and says, all fierce-like, that the Head wants to see me in his office. He reckons I'll be for it if I've cost us house points. I think I know what it's about. Yesterday, I threw a tennis ball in the playground; throwing anything is forbidden and, on top of that, it hit Foxey on the shoulder.

'Hold your hand out, Sambell,' the Head orders. 'Straight out, shoulder high.'

Whack, twice the cane hits my open hand.

'Now the other one. I'm disappointed with you boy. I trust this will be the first and last time this happens.'

My eyes water and my hands throb like mad, but I haven't drawn them back and I think, 'Ha, the conker game has paid off.'

In fact, it was the only time I was called into the office, as a pupil.

I'm standing on a football pitch in Palmer Park, shivering. We are in the middle of the 1943–1944 season, about to play against Blue Coat School. The temperature is about thirty-two degrees (Fahrenheit). I kick into the ground, it's solid. As captain of the school team I find I need to jolly up the others.

'Okay, so it's not good, the sooner we get started the sooner we'll warm up. It could be worse, at least there's no snow.'

'We're ready ref,' I call out.

'I'm not,' says one bright spark. 'I'm frozen to the ground.'

It's half-time and we are leading by two–nil. I've got a banging headache; the result of playing at centre half and heading out a near-solid leather ball several times. The ref says it's too risky to carry on and he's awarding the game to us. My side cheers. The opposition moans. But we are all glad to push off.

It's a wet Wednesday afternoon and off we go to Alfred Sutton swimming baths at Cemetery Junction – the Alfred Sutton of Suttons Seeds up on the London Road – to practise for the school swimming gala. I'm not a speed swimmer but I do have a good pair of lungs and some skill at diving and expect to do well in the plunge and on the springboard. In the month or so between footer and cricket, dry Wednesday afternoons are spent at the sports track in Palmer Park. I'm not a sprinter and practise the 880 yards and the long jump.

In my last two sports competitions I won the springboard for my House and in the final year I also won the 880 yards.

It is Wednesday, 10 February 1943. Harmy has arranged, for those who want to go, a visit to the Odeon to see a pre-release of *The Song of Bernadette*. It turns out to be a religious story about a girl who claims to have visions of the Virgin Mary. As we leave the cinema we are advised not to use Broad Street or Friar Street. We ignore it as a matter of course and, along with many others, Zeb and I aim for Broad Street.

'Why is there all this dust and smoke, Ron?' Zeb asks.

Before I can answer we get swallowed up in the crowds leaving the Vaudeville cinema. Grabbing hold of Zeb's jacket as we are carried along, I hear fire engine bells and see smoke pouring in from Friar Street. We were still in the Odeon when the bombs dropped and hadn't heard a thing.

We're sitting at the table as the six o'clock news tells us four 500 kilogram bombs hit Reading, presumably from a bomber hoping to hit the railway junction, which it missed.

What the pilot did hit was Simonds' Brewery and The People's Pantry – forty-one perished that day, twenty-six were in the restaurant, including one child. Had it not been Wednesday afternoon, an early closing day, the number of casualties would have been much higher. The People's Pantry was one of a chain of cheap restaurants set up by the government to cater for displaced people and to help eke out food rationing. It had a seating capacity of 100 and was manned by the Women's Voluntary Service. Ironically, it was the only eatery open that afternoon. Reading was bombed twenty times between 1940 and 1945.

Chapter 4

'The Journey' Among Other Things

On a late summer evening in 1945, Dinglebirdy and I step up to the front of the Collegiate Scout Troop. After three and a half years of scouting I'm the patrol leader of the Saxons and he is my second.

Dinglebirdy was Feliks Goldenfeld's nickname. We sat together in class and became close friends. His father owned an outfitters in Friar Street, where he helped out. He was a bit taller than me, rather podgy and he giggled a lot. He ran with arms flapping, head wagging and the nickname had stuck. He didn't seem to mind, probably because he was called a lot worse where he used to live. His family arrived in England from Austria in 1937.

As the Troop Leader unfurls flags and generally organises things my mind wanders and I'm back to when I first started at the school. Steero was taking morning assembly and arranged an afterschool meeting for new boys.

'Quieten down,' he ordered. 'I am the Scoutmaster here and I hope you will join the troop or the Cubs. We meet at the hall across the road on Thursday evenings.'

Before I left Old Ford I was in the Cubs at the Wesleyan Chapel; up shot my hand, 'I'd like to join the troop, Sir.' About a dozen others do the same. He grins and tells us that during Scout meetings we can call him Skip. It seems Mrs Harmy has a good stock of used uniforms and she kits us out after the meeting. I mention joining to Dinglebirdy and he said he'd like to have a go too.

Thinking on, to the turn of 1945, Skip raises his voice and beckons with his finger, 'Come over here, you two. Both of you need to think about gaining your King's Scout Award.' Dinglebirdy and I have our First Class badges and most of the qualifying tests for the Award just follow on. There is one final hurdle we need to prepare for and that's The Journey.

'I've been thinking about that,' Skip says. 'Since we'll be going to Port Quin in Cornwall for the troop's summer camp I reckon it will help enormously if there's an advance party to prepare the way by checking out the campsite, locating a dairy farmer willing to sell us milk, eggs and so on and a village with food shops. Doing that would be just the job for the Award.'

Little did we know what this 'job' would entail for two sixteen-year-olds.

I'd looked up the conditions for gaining the Award. Apart from the First Class proficiency badges we need to have attended two summer camps and at least three patrol camps, all under canvas. There are seven of us in the Saxons and I picture the trek cart loaded with all the stuff we need to spend a long weekend in a Niger tent. Six of us pushing, pulling and puffing, while Dinglebirdy marches along in front, throwing his stave up in the air like a drum major. You've got to hand it to him, he's nobody's fool.

At lunch one day I nudge him in the back and ask if he'd had any ideas for The Journey. He smirks as he tells me his Dad can rustle up two postman-type bikes which would do the job. They have big carrier baskets at the front and strong carriers at the back for saddle bags. That sounds great because for the test Skip's given us we will need to be kitted up for four or five days and be prepared to cover about 180 miles under our own steam. So, what we need now is the route.

'What we need now is a map,' Dinglebirdy chips in, stepping back as I try to cuff him.

This was easier said than done at the time. Because of the fear of German spies and invasion, all British maps had been taken off retail and library shelves, all signposts had been removed and milestones were either dug up or covered with cement. Peace had been declared on 8 May 1945 but it would be many months before road signs were restored.

I ask Skip about finding the way and it turns out he has an Ordnance Survey map of the North Cornwall Coast. Apart from that we have to work out the route for ourselves. It's geography books out, but the small-scale maps only show major towns and roads. We'll be travelling westward, south of Bristol for starters. Our navigating skills are going to be well and truly stretched.

Dinglebirdy and I struggle to get the bikes on to the road at the start of The Journey. Both carriers are piled high with an American bivvy tent, blankets, first aid bits and pieces, a Primus stove, a can of paraffin, a bottle of methylated spirits and spare bike parts. Cooking gear dangles from the handlebars. The saddlebags we'd cobbled together from army-issue backpacks and hung across the rear carriers were stuffed full of food, spare clothing, cleaning and washing stuff. Decked out as we were in scout uniforms on a hot summer morning the sweat poured off us.

Skip had forgotten to give us the OS map of North Cornwall so we had to make a detour out to Caversham to get it. From there we decided to cross back over the Thames at Mapledurham lock. By then it was getting on to midday and it was a relief to plonk ourselves down on the cool grass under the willows by the lock-keeper's cottage. The lockmaster, a woman actually, was at the down-stream-end of the lock, raising the sluices to let a pleasure paddle steamer through on its way to Abingdon. Her thick arms were turning the wheels that raised the sluice gates, for all the world like a helmsman fighting the wind on a clipper; the people on the boat were laughingly cheering her on. We had our Spam sandwiches there before heading towards Newbury, which we by-passed to the south.

We managed about twenty-five miles on the first day. Clapped-out as we were, as soon as a reasonable-looking farmhouse showed up we decided to call it a day. It was a dairy farm and we had no trouble getting milk, with bread and cheese thrown in for good measure. We pitched the bivvy on the grass behind the house. Sitting holding mugs of hot milk we worked out where the North Star was. Not as easy as you might think with millions of them looking like glittering holes in a jet-black sheet.

Next morning, we crawled out of the tent just as a boy from the farm arrived with steaming plates of bacon, eggs and fried bread. There's a bit of a problem, well it isn't really, Dinglebirdy has the eggs and I have the bacon. I ask him about the fried bread, shrugging his shoulders he takes a piece and gobbles it up, winking as it goes down. After striking camp we called at the farmhouse to thank the farmer's wife, then off on the next leg. This took us out on to Salisbury Plain, because we were determined to be able to claim that we'd camped at Stonehenge.

On the long undulating road across the Plain our one serious hiccup happens. I hadn't paid enough attention to stowing away the gear on my handlebars and the handle of a frying pan had snagged my front wheel. Before you could say Jack Robinson off I came, straight over the handlebars and into the ditch. There was no great damage to me but my front wheel was in a right mess. Dinglebirdy took it off and headed for the next village. Luckily, there was a garage there and the owner was able to replace the wrecked spokes. He drove Dinglebirdy, bike and wheel back to me and refused the money we offered as he waved goodbye and wished us good luck.

All this delayed us and we only just made it to Stonehenge before it got dark. The stones looked like massive rings of dominoes about to topple over. We'd learned a bit about Stonehenge in a geography lesson. There are several stories about how the monoliths got here, one in particular stuck with me. Some seven- or eight-thousand years ago a race of giants lived on Salisbury Plain and one night, when they were dancing in a circle holding hands, they were suddenly turned to stone. A bit creepy I felt, but sleep dead centre we did.

Then it rained the whole of the next day and we were so miserable we decided to look for a Bed and Breakfast in Taunton. After all, we both had two one-pound notes for emergencies. As it happened, one look at our soaked uniforms and the landlady put us up for free – and let us have a hot bath, all five inches of it.

Throughout this episode, and because of our Scout uniforms, we were shown a great deal of generosity and given much help. Our obligatory ration books were never needed and our one-pound notes remained intact. Although the war had ended, we saw nothing odd about the amount of bathwater. In 1942, the government had urged everyone to only use five inches in order to save energy, which was still adhered to by many. The landlady measured it with a ruler and suggested we bathed together.

On the third day our route took us south of Exmoor and we camped in the vicinity of Witheridge. On the fourth day we kept to the north of bleak Dartmoor and made it to Tintagel Castle by early evening. Having only ever experienced the hustle and bustle of the south coast's sandy beaches, at places like Margate and Southend, the ruggedness and solitude of the North Cornish coast rattled me. Waves crashed against massive cliffs and foamed over fallen rocks, the low wind-scratched clouds scuttled inland. The bent-over trees, brambles and heathers were clinging desperately to thin soil. Tent pegs were useless, we tied the guy ropes to heavy pieces of rock. I made sketches of the ruins and the view across the Celtic Sea for inclusion in our report.

Since crossing the Tamar into Cornwall at Polson Bridge, Launceston, we were able to make use of Skip's map. Port Quin was just ten or so miles further south west and we got there about noon on the fifth day. To say I was knocked sideways is putting it mildly. There was nothing there – no houses, no people, no shops, no boats – just a couple of rundown buildings and an enormous seabird gliding towards us.

I turned to Dinglebirdy and groaned, 'Please tell me we're in the wrong place.'

He shook his head, threw his hands up in the air and laughed his head off. We went back up the lane to a farmhouse we'd passed.

The chap there took one look at us and said, 'You're late, never mind, come on in and I'll explain.'

It turned out Skip had been in touch with him and we were expected, on the assumption we'd use our brains. We all went back down to the beach and he showed us where we could camp. He'd be providing milk, eggs, cheese and some meat in the form of chickens and rabbits. A place called Trelights was the nearest for shops. That afternoon the troop arrived.

In the 1600s Port Quin – Cornish Porth Gwynn or White Cove – was a thriving village of over 300 fishermen, antimony and lead miners and families. Over the next 200 years the herring catches the villagers relied on dwindled and the 1841 census shows there were ninety-four souls and twenty-three inhabited cottages. Legend says that all the fishermen were drowned in a storm at sea because they sailed out on a Sunday, thus breaking the Sabbath. Reality is the villagers migrated to places like Port Isaac and Polzeath. There has since been some redevelopment by the National Trust in the form of holiday lets.

I enjoy Louis MacNeice's poetry and when I sought a title for this poem his phrase, 'It's no go the merrygoround' came to mind – it's the first line of his poem 'Bagpipe Music'. I have used it rather drily.

Merry Go Round

Port Quin tucked snug beyond Cornish cliffs
Safe from the gales and raging seas
Gaping razor shells in whispering sand
Swathes of kelp on a virgin beach~
Fishing smacks, crab boats, nets and creels
Storm-hardened mariners cull their catches
As wheeling scavengers diving and scolding
Snatch at the offal, twisting, stalling
Expectant fisherwives chatter on the quay
Jealously eyeing the threatening swell
Hands clutching baskets dependent on the haul
The whole interlinked in a perennial swirl.

Dinglebirdy jogs me and I look up to see Skip coming towards us. In his right hand he is holding the two King's Scout Badges we have earned. He tells the troop about The Journey and our efforts to search out a suitable campsite for our recent summer camp. He shakes us both by the left hand and gives us the awards. These badges, which are larger than the others sewn on our shirts, are on display for the next meeting.

> *When the troop started clapping as we accepted our awards I spontaneously grabbed Feliks' arm and pulled him close. Ethnicity was not in my vocabulary, this was simply a gesture to a close school friend. My early years were spent in and around the East End Jewish Community: Mrs Segal, the hairdresser next door; the Konraths and their daughter Josephine had a bakery down the road and Jo would regularly deliver bread to us on the way to school; Dr Lightstone, our GP; Dad's barbers in Whitechapel Road where he would take me – I remember the hot towels when someone was shaved with an open razor and the lively interchange of news and chat between all present. These were friends of my parents, and to me simply 'people'.*

It is November 1945 and I'm on my way to Maidenhead to sit my Cambridge School Certificate exams. I'm not worried about them, the school has done a good job of teaching me. Out of the train window I see some of the fortifications built at the time the war began.

That's all over now, but the celebrations on Victory in Europe – VE Day, 8 May – are still fresh in my mind. We had a whale of a time. Somehow, people had found all sorts of food and had set up tables in the streets for children's parties in the afternoon. Everywhere was festooned with Union Jacks and decorations fluttered in a smart breeze.

Later on, I caught up with a group of friends and we all set off for the town centre. It was packed with people. A couple of local bands had set up, one in Broad Street and another in Friar Street and they were going at it hammer and tongs, beating out the latest bebopping music. Because of the Bomber Command units and the Americans based at Greenham Common, there were many RAF crews and GIs around and the girls were showing off jitterbugging with them. Pauline Wilson was in my group; since the library thing we had met on and off, going to parties, dances and plane-spotting club evenings.

As a seventeen-year-old I had the usual interest in girls and counted a few as social friends, so it didn't mean too much to me when she asked me to have a dance. Perhaps it was the excitement of the day but I began to realise we were taking stock of each other and splitting off from our group. That evening Paul came home with me and we sat in the front room on the sofa listening to some of my records. One of my favourites was 'Brown Eyes Why Are you Blue' sung by Layton and Johnstone. As I stood up to rewind the gramophone she pulled me back and out of the blue kissed me. We are now going out together regularly.

Dinglebirdy is coming out of the Head's office as I wait my turn. He's done okay, having got six credits, including business studies. He's got a big smile on his face and tells me that, along with his family, he is migrating to Canada sometime soon.

As a leaver on this last day at the Collegiate, the Head calls me to come in, shakes my hand, gives me my results and a character reference. Education-wise I find I've done well, gaining a matriculation exemption from university, with distinctions in English and English literature and credits in maths, history, geography and religious studies. My head swells. Thanking him, I turn to leave but he asks me what my plans are. Frankly, I'm not sure and I explain that the two years spent on a dairy farm made me think about becoming a vet. There would be an entrance exam but the problem was I had no science and I'd need another language, all at School Certificate credit level to get into vet college.

He rubs his chin and agrees it's rather a long shot, but I'm a quick learner with a retentive memory and I should think seriously about giving it a try. He advises, 'Take two-year crash courses at a polytechnic and go from there.'

We are going back home, to Old Ford I mean. All our goodbyes have been said. Mine and Paul's was a tearful one. After some consoling – okay, okay, after some kissing and cuddling – we promise to stay in touch.

Standing at the front window of number 12 I see Dad drawing up and I give Gran a call. Zeb doesn't move from where he's sitting on the sofa. I can't blame him, I'm worried too. September 1939 is such a long time ago I really and truly don't know what we should do. It sorts itself out when Gran tells us to move ourselves and get outside. We are all at the gate as he gets out. Amazing, I'm taller than Dad by several inches and his hair is grey. Realising that the remoteness I feel has always been there, I let out my held-in breath and take a step forward, smiling. Dad breaks the ice by commenting on my height and I ask after Mum. Gran tells him to come on in and rest himself with the obligatory cup of tea, adding that we are ready for the off. Zeb and I load up the car and within the hour we are on our way.

Gran is sitting up front, us two in the back. After one or two stabs at chatting we all fall silent for the rest of the journey. Where are we at? Or, more to the point, where am I at? How have the last six years left the balance sheet of my life so far? This question is buzzing around in my head as I stare out of the car window. Mum and Dad, Zeb and me, we are no longer a family unit. Oh yes, I'm sure there will always be love and care, but sadly, cohesion and communication are going to be absent; as are family get-togethers, meeting up with mates and the like. Being winkled out of my back-street environment has alerted me to a world in which I think I have the tools to shape my future, which far outweighs the loss of my earlier way of life. I sit back, close my eyes and decide my balance sheet is firmly in the black.

Chapter 5

Lost and Found — the In-Between Years

The extent of the Blitz damage is mind-boggling as we go through Hammersmith, Central London and into the East End. In Old Ford it hits me hard as I see the station is badly damaged, as are the houses in Wendon Street, round the corner from us and parallel with the railway. I knew so many of the families living there. Lefevre Road is now gone, flattened by a V2 version of the doodlebug. How the line of four shops and the house on the corner survived with just a few blown-in windows beats me.

One of the customers must have told Mum we'd arrived since she appears at the door, wiping her hands on her apron. Again, I don't know how to deal with this, but she does as she runs across the pavement, puts her arms round us, hugs and kisses to our great embarrassment. Dad knocks it on the head by telling us to get the car unloaded. Armed with suitcases, Zeb and I go inside to be faced with great cheering and clapping and further discomfort, as Ruby rushes forward and plonks a smacker on my face. Zeb is so embarrassed he turns his flushing cheeks away and tells her not to be so daft. I ask after Dot, Iris, Maggie and Flo. Ruby looks down as she tells me Dot died in an air raid. I hug her and tell her how sorry I am, Mum hadn't told us. Iris got married but still works for us. Flo and Maggie are safe and sound though, and Annie, who took Dot's place she tells me, has become another great favourite with the customers.

We all go through to the kitchen. It's so much smaller than I remember. My safe place under the cutlery shelves is far too cramped to crawl into. Flo looks at us a bit cautious, so, with a big smile on my face, I promptly put my arms round her and ask how she's doing. That flusters her and, before I can blink, she bursts out crying. Ruby tells her to behave and sits her down in Grandma's old corner.

'But he's grown up,' Flo manages to blurt out as she starts off again.

The kerfuffle dies down and Zeb and I finish unloading the car and lug the cases, boxes and bags upstairs to the parlour. Mum follows us and tells us which rooms to put them in. Clearly, Zeb and I can no longer share a bed, so she has given the bedroom to him while I'll be sleeping on a bed-settee in here. I feel a bit miffed about it but let it go. Gran has her old room back, which is behind Mum and Dad's double bedroom.

By this time the business has closed for the day and Dad is back from garaging the car in Ruston Street. The evening meal is about ready and Mum asks me to lay a table in the toff's dining room. Dad raises his eyebrows at this but Mum is having none of it and tells me to carry on. She is cooking old favourites: individual steamed steak and kidney puds, mash and greens, followed by apple pie and custard. Owning a restaurant has some very definite advantages during rationing.

The need for ration books eventually came to an end on 4 July 1954, when the control of meat and bacon was lifted, nine years after peace was proclaimed.

Going upstairs to get my pullover I see that things like fags, fag papers, cigars and baccy are still stocked on the same shelves. I remember swapping my doubled ciggy cards for ones I hadn't got from the stock packets. I'd lie when accused of doing it, but I wasn't very good at it because I got caught out regularly.

I've slept well and come-to as Dad sticks his head round the door and says we need to have a chat after breakfast. Later, as I follow him through to the dining room, Zeb trails behind, but Dad turns and tells him we're busy and to hop it. We sit and he makes no bones about asking me what comes next. All this bossing about is new and causes me to shuffle a bit. As I tell him what I would like to do and how I might get there, his face seems to close up. In reply, he thinks it's going to cost a lot of money in college fees and did I really think I could do it. I tell him I'm not sure and that it will, in any case, depend on passing some science exams and learning a foreign language, while he lights a fag, coughs and scratches his head. With half-closed eyes he reckons we're looking at four or five years before I qualify. More like six, I think.

He becomes a bit more constructive, saying that the extra qualifications in sciences and a language will broaden my job possibilities and he will support me while I study for these and then look at what comes next. I'm more than happy to go along with this and set about enrolling at Regent Street Polytechnic and a privately run language school. Nine months of savage tuition in French is enough to get me a credit at School Certificate level. Another year on and I get passes in physics, chemistry and biology at the Poly. I'd never crammed so hard in all my schooldays.

At nineteen I go for an interview at the Royal Veterinary College in St. Pancras. Harmy had me taped that's for sure and Dad has gone along with my needs. The interview is oriented more toward background and personality. It is successful, despite my origins and my noticeable cockney accent. My matriculation exemption means I don't have to sit the entrance exam. The first year turns into a nightmare. I really had needed to attain at least credit-level grades, particularly in biology, and I only just scrape through the exams. The second year involves in-depth organic chemistry and anatomy – dissection of dead dogs and rabbits and live treatment of dray horses with sinus problems in the brewery down the road. Enough is enough. Although I just manage to pass, I call it quits. Dad blows his top but realises he's throwing good money down the drain. What now? My national service is well overdue since for most men enlistment is at age eighteen and here am I approaching twenty-one; as an ex-student, call-up looms.

Thinking back to when we first came home in 1945 I marvel at what has taken place. First, Mum and Dad decided to buy a house. I can't follow their reasoning, but they move out to Gidea Park in 1946, to a detached house named Walnut Tree Cottage. Gidea Park is on the Great Eastern Line into Liverpool Street Station. I stay on in Old Ford on weekdays, since it's more convenient during my two years of cramming and my vet college years, 1945–1949.

In 1957 Mum and Dad made another move, to 28 Bedford Road in South Woodford. This is in anticipation of the row of shops in Old Ford being compulsorily purchased by the council. Mum bought a restaurant in Wood Green with the proceeds.

Apart from a few letters – from me that is, since she wrote regularly – my contact with Paul is minimal. Dad had held on to 12 Hamilton Road in Reading and let it to the Prestons. I know their children, Patrick and Hilary, from school days and decide to contact them to see if I could stay for a weekend. They are happy to oblige and in the approach to Christmas 1947 I visit. I tell them my main reason for coming is to see Paul. What I haven't realised is, I may have blown it.

I had phoned Paul to say I was coming, but she'd already arranged a date with an old school flame and wasn't prepared to break it. She'd meet up with me on the Sunday. I knew John from the schooldays' gang. He's a year older than me, a well-built good looker and still into Sea Scouting. I just knew he must be in with a chance. However, my worries disappear as soon as we meet. The attraction between us is as strong as before and when we stop for breath we're both in the clouds. I use this arrangement several times; being a lousy letter writer, giving voice is the only way I can interact with and convey my love for her.

On one of these weekends we take an evening walk through White Knights Park. It was gloomy so we didn't realise until it was too late that we were treading on frogs, dozens and dozens of them crossing the path on some form of migration. With an ear-splitting shriek Paul was on my back in a flash. There was no alternative, I had to keep going for another ten yards, it left me shaking and locked in a stranglehold.

During the winter of 1948 I tell Mum and Dad about Paul and ask, will it be okay for her to come and visit? Neither of them shows much enthusiasm, so I let things drift until I tell them of my decision to leave vet college, when I speak to them about it again. I emphasise our love for each other and that we'd been seriously dating for about twelve months. They close up as I tell them I'd been back to Reading several times. Dad rants a bit, saying it was the cause of my vet college difficulties. Voices are raised as we argue and it underlines the lack of understanding in the family. Finally, Mum looks at the situation a bit more pragmatically and agrees it makes sense to meet Paul before matters get out of hand. I lose no time inviting her and give her directions to get to Gidea Park, which become unnecessary when Mum suggests she be picked up at Paddington. The visit is an almighty disaster; I don't think I shall ever be able to forgive my parents for making Paul feel so unwelcome.

I see a life lesson in all this and promise myself I will always make the future friends and spouses of my children and grandchildren unreservedly welcome.

Chapter 6

The Best Half

Pauline Patricia Wilson, who preferred to be called Paul, was born on 3 May1930. Full grown, she was about five feet, four inches tall, pretty and very nicely proportioned, blue-eyed and with fair wavy hair. Until she was evacuated in September 1939, she and her mother, Patricia Anne Wilson, lived with her Mum's sister Ellen Maud and husband, Henry Augustus Wilkinson. Her uncle-in-law was the licensee of the Lord Nelson pub on the corner of Trafalgar Road, a stone's throw from the Elephant and Castle. Ellen and Henry had two children, Peggy and Ron – Paul came to look on them as sister and brother.

In 1939 she was evacuated to Wells where, because of her blossoming singing voice, she formed a friendship with the choir and the lady organist at Wells Cathedral. The family she lived with didn't appreciate having a child of nine being allocated to them and requested she be housed elsewhere, so after a year she was rehoused in Yeovil. The Blitz was in full fury by then and the Lord Nelson was badly damaged. Henry leased a place in Reading, a three-storey, double-fronted house, 20 Bulmershe Road, next to Hamilton Road. Paul's Aunt Ellen turned her home into a boarding house. Henry worked as a 'canvasser for house furnisher'. He passed away, apparently without the regret of his family, on 15 November 1941. Paul's Mum had moved with them to Reading and once things settled down they brought Paul to live with them. Sadly, Paul's Mum died soon after and Aunt Ellen, or Mops to her family, took Paul under her wing.

As you are already aware, our paths crossed that knickers day in 1942. Paul was like no girl I've met since; full of backchat, expert at double entendre and outright tomfoolery. If she's embarrassed me once, she's done it many times.

Nevertheless, I love her for it. Despite my social reticence she pulled me along in her wake, never putting me to one side. She was a good athlete; at fifteen she was playing hockey for the Berkshire Scarlet Runners. The day came when her hockey stick got caught up in the front wheel of her bike and she went flying, it took her many weeks to recover.

School leaving age was still fourteen, unless you went on to secondary school, which was available at Alfred Suttons and Paul stayed put. Her strong-mindedness soon became apparent in terms of what she was prepared to be taught. In those days, the emphasis for girls was on becoming equipped with skills suitable for office and secretarial positions. After a few weeks of this Paul threw a typewriter on the floor and walked out. Despite this short sortie in commerce she became very adept on the typewriter, her fingers flitting over the keys like finches in a thistle patch.

Fortunately, science subjects were available. I don't know how she managed it, but she was allowed to stay on to do sciences and maths. She absorbed all of it, particularly biology, and at rising seventeen she left. It so happened that three of Mops' lodgers worked as professionals at the National Institute for Research in Dairying (NIRD) at Shinfield. Paul's immediate step was to solicit their aid in gaining a post at NIRD. The emphasis there was on microbiology and immunology and she studied part time as a bacteriology laboratory assistant. She gloried in her five years there, then I put a spoke in it.

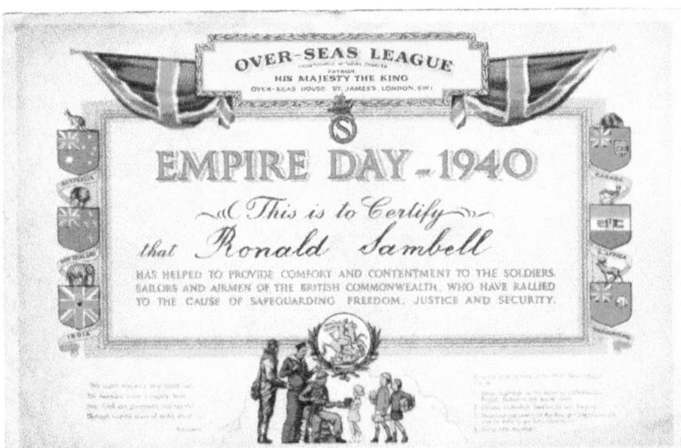

1940 Empire Day. Presented to us by a bigwig from Reading for our knitted and newspaper contributions to the war effort

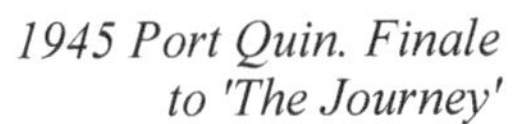

1945 Port Quin. Finale to 'The Journey'

1944–45 Football Captain at Reading Collegiate School. Fourth in from left, back row

1948 Paul, my favourite photo
when ever I think back

1947 Paul at the National Institute
for Research in Dairying

1946 Pool petrol ration.
8 old pence per gallon
(about 1p/litre)

1946 Relaxing with Paul at No. 20 Bulmershe Road

1949 Making the most of our honeymoon!

1949 Sept 3rd. A very happily married Paul and Ron.

1949 We couldn't get away fast enough.

1950 Me showing off my military status.

1951 Patricia's christening day

1964 A proud me with Lindsay at College Road

1966 Official programme for the final

1966 GB mascot

1966 What a day to remember!

Chapter 7

Heaven Help Us

My call-up letter came the day before my twenty-first birthday, telling me to report at North Camp near Aldershot on Wednesday 15 September 1949. My first actions are to let Mum and Dad know and then to phone Mops. I tell her the news and ask her to get Paul to ring me when she gets home. After a brief chat in the evening we agree it would be much better if I visit. Paul comes straight to the point about the situation when I get there and asks me what comes next. This is an easy one. I promptly ask her to marry me. She straightway calls out to Mops that I've just proposed. To which comes the reply, 'It's about time. I hope you said yes.'

The following day I break the news to Mum, who looks at me as if I've gone starkers. She thinks it best to tell Dad herself when he gets home that evening. I shrug. Whatever, it's not going to change anything, I was sure we would get married sometime so why not now? I had been pleasantly surprised to learn that a private's marriage allowance was 14/- per week. It will make a big difference, since my pay as a private will be £2/8/- per week.

Three weeks fly by. One of Mops' lodgers, Helen (a beautiful blonde dress-maker whose husband Ben had survived the war fighting with the Polish resistance), has come up with a white silk parachute which she transforms into a stunning wedding dress. On the morning of our marriage, Mops' son Ron Wilkinson, my best man, points at my shoes and reckons they look very shabby and I need to get new ones. With just two hours to go we traipse off into town and do the necessary.

Our marriage takes place in St. Bartholomew's Church, Palmer Park, at 1.30 pm on Saturday 3 September 1949. Paul, looking radiantly lovely, is given away by my Dad – a moment of inspiration on her part. Hilary Preston, as one of Paul's closest mates, is her bridesmaid. Her matron of honour is her cousin, Peggy Goodenough, Mops' now-married daughter. The numbers attending surprises me, although it shouldn't have. Apart from immediate families and the six lodgers, there are many staffers from NIRD, a cluster of Paul's friends, the Preston family and her friends the Huggins, from up the road. Some wit observes it's the tenth anniversary of the day war broke out and suspects it's about to happen again. Mops conjures up a cracking wedding breakfast, considering the food rationing, and then we are off and away.

Off and away is a week at Pagham Beach, not far from Bognor Regis. We leave from Reading on Southern Railway to a great send-off, with a 'JUST MARRIED' placard in the window of our third-class compartment. When the guard sees what's happening, he yanks us out and, with a big grin, puts us in an unoccupied first-class compartment with a 'Do Not Disturb' notice.

One of Mops' wedding presents was to rent a caravan for us, called Eve, just off the beach at Pagham. We take our bikes with us and more food than we know what to do with. This has to be the happiest and cheapest wedding and honeymoon on record.

Back to the nitty gritty. Mops had agreed we could rent her two converted attic rooms as a flat and I bring what possessions I have from Gidea Park to Reading. With only five days before I join up, Paul and I go to ground and while away the time making the flat look a bit more like home.

She comes with me to North Camp. We are able to get there direct from Reading by train, a distance of sixteen miles. There is no problem finding our way to the reporting office as I am one among dozens. There is quite a cheer as we kiss our farewells since there are very few with someone to say goodbye to. I find myself in the unusual position of being three years older than most of the conscripts – and married to boot.

After being registered we are arbitrarily divided up into units of twenty and marched off to the supply depot, where we get kitted out in uniforms, underwear, socks, boots, fatigues, eating irons, mess tins, towels, soap, sheets and cleaning materials, a duffle bag to put it all in and a backpack. Next stop, a Nissen barrack room with ten beds down either side, each has a rolled-up mattress and two neatly folded blankets. There are two bathrooms at the far end. We are told to get changed into uniform, make our beds and wait. Eventually a corporal turns up and orders us to get our eating gear and fall-in outside; it's gone 2 pm so there is no holding back.

Later that afternoon my turn comes to have a headquarters interview. Having already filled in a form giving my education and status the three officers on the board are well briefed. I'm asked about my attitude regarding infantry placement and active duty. It appears I am potential for commissioned officer. Would I like to attend an interview to that effect – with the proviso that it would mean signing on for extended service? I decline this way forward and remind them I'm married and would much prefer to be in a non-combatant regiment, such as the Royal Veterinary Corps or Royal Army Medical Corps, with a view to learning a skill that I could make use of when demobilised. After some discussion, they assign me to the Royal Army Dental Corps because the RADC has a training school for support staff on-site and, because I'm married, they see no reason to post me far afield.

The next six weeks are spent doing basic training; learning the orders and manoeuvres on a parade ground, route marching, shooting on a rifle range and gaining basic outdoor survival experience. I am tickled pink to find I'm the only one in a unit of six who has any idea how to look out for his self. The majority of the intake then transfer to Army regiments to learn warfare and technical skills. I find myself in a smaller Nissen with seven others, all doing the same as me. Within a short time I become friendly with a Scotsman, Alasdair. He is a committed pacifist, proud and calm with it. He decides to take up the offer of staying put for four years to learn the trade of Dental Technician. This I also say no to and train instead as a Dental Operating Room Assistant – the acronym results in lots of ribaldry on the parade ground. However, the teeth of many of these eighteen-year-olds are in a parlous state, which gives us lots of opportunities to get our own back in the treatment room.

Being in a non-combatant training unit means the time spent in class is an hour or two a day at most. We are on a long rota list for other duties, such as square bashing, night patrol, officers' mess support and keep-fit. This means that weekends can be made duty free by doing swaps with the others. During basic training there is no leave, but yours truly is off like a greyhound on the first weekend after it finishes. My first home-coming is magical, hours are spent just holding tight and it takes every ounce of willpower to say goodbye. I cycle back to North Camp in time to get to roll call on the Monday. This twenty-mile journey by road is repeated close to a hundred times over the next two years.

I have no trouble with the dental course, it is mostly common sense and the technicalities are well within my clinical knowledge. I pass all the tests and am upgraded to DORA 2*, Grade II Private. Within ten months I am rated DORA 4* and a Group A Grade I Private – which means I get additional pay of 17/6 (87.5p) a week. Generally this leads to a promotion to lance corporal, but that is out of the question for a non-combatant national service soldier. I'm not concerned, in fact, it would curtail my visits home.

In June 1950 Paul gives me the disconcerting news that she is pregnant. We've never discussed starting a family and I am immediately out of my depth. Paul looks at me a bit sideways as she tries to hide what is clearly a 'fly me to the moon' moment for her. I take her in my arms and attempt to be practical. She tells me to shush and take a deep breath. Everything is going to be well in hand. I will be out of the Army the following autumn and, in the meantime, Mops will be at her side. Am I okay with this? Once I get over the shock, I can't stop saying 'Yes'. We go down to find Mops and a threesome of hugs and kisses happens as Mops tells Paul she knew I'd be over the moon too. I can't help being on tenterhooks though. Back at the barracks I report the news to my commanding officer and I can't keep it in from Jock. Net result, everyone knows by the next day. I take a deal of ribbing and no mistake.

It's a bonus to learn that the arrival of a child carries an allowance of 12/6 (62.5p) a week. Patricia Anne Sambell is born on 14 February 1951, St Valentine's Day. I get the statutory five days leave and arrive home on the sixteenth. Apparently all had gone smoothly. Patricia was born at home; it was like shelling peas, the midwife says cheerily.

Our senior non-commissioned officer is a staff sergeant, a war veteran who had served with the Royal Berkshire Regiment, a lean, stern-faced ramrod, well over six feet tall. He is responsible for ensuring we act in a soldierly manner. Put a foot wrong on the parade ground and we are made to rectify the mistake over and over. If your boots don't have a glassy polish or your uniform trousers lack a razor crease, it's an extra night-guard duty for a week. If a bed is slipshod in the making, it's off to the officers' mess for domestic duties. My dismay when he barks out my name and orders me to follow him, is accompanied by several snide remarks from my mates. Without saying another word he leads me away from the barracks to the married quarters. Perhaps I'm in for some housekeeping?

'This is where I live,' he states and introduces me to his wife Margaret, as his face dissolves into a warm smile. Without more ado, they take my hands and shake them while congratulating me on becoming a father and inviting me to stay for dinner. As the saying goes, 'you shouldn't judge a book by its cover'.

Phew, that was a close one, but it is effectively closure to a strange period in my life. Things chunter along until my demobilisation in September 1951.

Chapter 8

The Holes Get Bigger or Disappear

The arrival of the fair, curly headed Trish puts paid to Paul's job at NIRD. My adieu to the RADC scuppers any income. Does it worry us? Not one jot. Neither of us has the spending bug and we have put a few quid in the piggy bank. In the morning I'll be off to the Labour Exchange.

What a depressing place that turns out to be. I'm tempted to leave the door open, the cigarette fug makes me gasp. I see a dozen would-be workers slumped on the backless wooden benches lining the dark-brown walls. A sign tells me if I'm a new applicant to fill in one of the forms below – arrow pointing down – and hand it in. With that done, I join the unemployed.

I'm reading a leaflet about benefits when my name is called. Sitting at the desk is a balding man with steel-rimmed glasses, which he pushes back on his bulging nose as I stand before him. Scratching his thinning ginger hair with his pencil he informs me that being a DORA in the Army is not going to help one bit, men in that job are rarer than hens' teeth since it's done by women in Civvy Street. He glances at my answers on the form and doesn't think there's anything going at this time. I get the impression he thinks I'm a lost cause. He writes on a printed card, stamps it and tells me to take it to the Pensions and Benefits Offices.

These offices are in Bridge Street. The clerk gives me another form to complete. As a married man with a wife and one child I am given a voucher for £3/17/-. It appears this is just below the average weekly wage for an unskilled worker. I shall receive a book of these vouchers in due course, to be cashed at the Post Office on a weekly basis until I find a job. Considering that I no longer get the financial advantages of Army life – free bed, board and clothing – we are losing out.

On my fifth daily visit to the Labour Exchange I jump at the chance to work as a porter at the Royal Berkshire Hospital for £3/19/- a week. I am advised to come back once a week to see if jobs with better prospects have turned up.

A clerk in the hospital's admin office issues me with overalls and a peaked hat with RBH stamped on them. The foreman I report to runs through my round of duties, followed by a Cook's tour of the hospital. There are about thirty of us and our jobs range from checking-in and storing deliveries, to pushing patients on gurneys from A to B.

The foreman tells me, 'It's not all wards though. There are day-care facilities, like this one here.' He pulls back a door. 'The ophthalmic glaucoma unit.'

I peek to see several people sitting patiently but comfortably in armchairs. Nurses and practitioners glide around the reassuring environment. At regular intervals on the white walls there is a repeating plaque, 'Clean hands? Please use the facilities or ask for help'. I wonder if anyone ever does, ask for help that is.

After a week of this I eagerly revisit the Labour Exchange. I pass the time looking at the dozen or so notices stuck at random on the walls. Most relate to benefits and welfare and are lost among several large, fading and peeling posters dating back to the war. They are of the slogan variety: 'Walls Have Ears', 'Dig for Victory', 'Careless Talk Costs Lives', 'Be like Dad – Keep Mum'.

Whoosh, the door flies open and in breezes an oddball wearing a smart grey suit. No taller than me, arms swinging, feet spread at 2 o'clock and black hair plastered down. Before he gets across the room he's asking the clerk if there is anything for him today. Cheeky chappy comes to mind. Steel Rims looks up and shakes his head but then tells him to hold on, there is a possible. He riffles through the applications then calls out my name and says to come over. While the man in grey gives me a good look, the clerk trots out my details. There is a moment's hesitation before Cheeky Chappy tells the clerk he is taking me into the back room for a chat.

No introduction, straight into questioning me: can I use a slide rule, a chemical balance, a microscope? It's 'Yes' to all those and more. He sits down with a frown. It seems there's a problem; I'm too qualified for the job he can offer me.

'It's why I hesitated out there,' he mutters.

He's looking for what he calls a laboratory attendant as a pair of hands in a metallurgical laboratory, whereas my qualifications would start me off as a scientific assistant (SA), for which he hasn't got an immediate vacancy. The wage, he calls it a salary, would be about £3/16/- a week, with the carrot that as soon as a chance did occur he would recommend me for promotion to SA. My mind is in a bit of a whirl. Our piggy bank is almost empty and the wage is a pound short of what we need right now. I want time to think and ask him if the lab is local. He shakes his head and says it's about twenty miles away. I boggle, but he quickly adds that free transport is provided. He stands up and stares at me. Gulp. I make a spontaneous response, 'I'll take the job.' Then he does say he's Dr Clarke, a scientist working for the government, and that he can't tell me more until I have been cleared for security purposes, which will take a few weeks. We go back to the counter where he gives a thumbs-up to the clerk and leaves. I give a wry grin and thank Steel Rims for pointing me out.

I can't wait to get home, certain sure there's nothing in my background that will label me unsuitable. When I get back Mops is on her knees scrubbing the front step. She gives me a slap on the backside as I break the news. Then it's two at a time up to the flat. I take a deep breath and open the door. Paul looks up from some ironing. Trish is asleep in her cot. I flop down on the sofa with a big sigh and ask her to come by me. She looks concerned and wants to know what's wrong as she sits. I can't help smiling and look away. I airily tell her nothing is and I've only landed another job. She whacks the back of my head and jumps on me. 'Tell, tell, tell.' She gives me an ear-bashing, but the sparkle in her eyes makes the shortfall in income a worthwhile sacrifice.

Placing her hands either side of my face and looking earnestly into my eyes she says she also has a bit of news, she is expecting. My hands cover hers as I struggle to cope with this change of tack. 'When? Are you okay with it?' are my first reactions. Then it sinks in. I yank her up and whirl her around in a show of pleasure. The baby is due next June and I am the first to hear the news. We tell Mops and then take stock of our situation. It's pretty clear we need to think about somewhere else to live, especially since Mops has been hinting she has her own reasons for wanting us to give up the flat. We look at the pros and cons and my prospects. Assuming I get the job with Dr Clarke, we reckon I need to be doing a bit of moonlighting to supplement my earnings.

As I'm moving beds in a ward I'm confronted by a black-suited heavyweight who shows me a pass that identifies him as a government employee with the Home Office. He is quiet as he leads the way to an unoccupied room. His stony features break into a grin and, shaking my hand, he gives me the good news that I've passed the security check and I'm free to accept Dr Clarke's offer. He cautions me about the sensitivity of the information I shall be privy to and I sign something called The Official Secrets Act. I'm stumped, what on earth am I letting myself in for? I ask him where I'll be working and what all the secrecy is about. It seems everything will come to light when I receive a letter of confirmation in which my grade, pay, job description and its location will be made clear. Then he wishes me good luck, shakes my hand again, and leaves.

Paul waves a letter under my nose when I get home from the hospital a few days later, it's an 'On His Majesty's Service' envelope. How she resisted opening it I can't think, but now she is crowding me and telling me to be quick. I open it carefully and fish out a letter from the Ministry of Supply. It informs me I will be a Laboratory Attendant Grade 9 with a starting salary, that word again, of £240 per annum (£4/6/- a week). I am to report to the main gate at the Atomic Energy Research Establishment (AERE), Harwell, on Monday 15 November 1951. If I wish, I can use the coach leaving Cemetery Junction at 0740 – show this letter to the bus steward, it will act as a pass. A feeling of relief goes through me and I let out a deep breath. The weekly wage is more than I'd dared hope. Paul puts her arms round me, gives my hair a tussle and insists we celebrate.
In the meantime, we've been giving serious thought to moving out. We think the only way we'll be able to do anything at all is to take on a large house and

do what Mops does, only in our case we'd take Reading University students, with the resulting breaks when they are down. There are three choices: rent, lease or buy. We ask Dad for his advice and decide to buy, since he is prepared to lend us a deposit. The simplest way is to take out a mortgage in the form of an endowment policy with profits. Again, Dad does the proverbial and finds that Britannic Assurance are prepared to lend us £2,000 on a fixed 2% interest rate over the twenty-year lifetime of a with-profits policy to the same amount, with Dad as guarantor. Considering I'm effectively between jobs it is something of a miracle.

Eagerly we go house-hunting, only to find it's a deflating exercise. This one's too costly, this one's too small, until a friend tells us that one is about to go on the market in College Road, the next road up from Mops' and opposite Palmer Park gates. Number 23 is a semi-detached house with three reception rooms, a dining room, a large kitchen, four bedrooms and a loft. The owners want £2,600 for it. After a bit of bargaining we get it down to £2,500. It's a close shave, Dad can only lend us £500. Within a month, we are the new owners of an unfurnished house.

Chapter 9

Four Strands and Consolidation

It is inevitable with a large family and a lot going on for there to be several timelines in parallel, viz. Me, Paul, the Children and Others. I'm allowing the story to run loose until the timelines intertwine, coalesce or conclude.

I'm on the Harwell coach. It's Army issue: slatted wooden seats, no heating, camouflage green all over and noisy. It takes us out of Reading on the Bath Road through Pangbourne to Streatley, then across the Berkshire Downs, where we turn left at Rowstock Corner, Harwell, and after a mile or so reach AERE.

The gates are manned by two armed security guards and one of them ushers me into the guard house. My identity is verified and I'm escorted to the Administration Block, Building 329 – every building is known by a number. I'm photographed, fingerprinted, reminded about secrecy, issued with a pass, two white coats , a close-fitting peaked hat and a pair of industrial shoes. Then a clerk leads me to Building 354, Metallurgy Division, a good quarter of a mile away.

My Section Leader, Dr Alan Blainey, is not immediately available and I meet Henry Lloyd, second in line. Harry, a forthright but friendly northerner, spends some time explaining the nature of the group's work and how I will fit in. My job entails general housekeeping in three laboratories, ordering and collecting stores and equipment and providing a pair of hands where needed. There is a steep learning curve in the last task, where it is routine to use equipment such as Geiger counters, high-temperature furnaces and vacuum chambers.

AERE was founded by Sir John Cockcroft in 1946 to provide research support for the nuclear energy industry and the nuclear defence arm of the military. Harwell, a decommissioned RAF airfield, was requisitioned as a site. It was funded by the Ministry of Supply and Sir John was its first Director. In 1954, AERE was incorporated into the United Kingdom Atomic Energy Authority, an autonomous government agency.

This first day flies by and as I get off the coach at 5.30 pm I feel a bit shaky, my mind numbed by it all. Paul is full of questions as we sit at the dining table with Mops and her lodgers. Recapping my day settles my mind and I begin to realise what an incredible opportunity is opening up for me.

My job at Harwell begins to shape up. I help Alan with a project of his own that involves a new plastic, polytetrafluoroethylene (PTFE), known commercially as Fluon or Teflon, classified as the most slippery material known. The nuclear industry wants information on the effects of gamma radiation and heat on its plasticity and friction properties. The radiation results are classified, the thermal results are not, and the latter work is presented along with other papers by Alan at a Chemical Engineering Conference. It makes my day to read my first formal acknowledgement in the published proceedings.

Freddie Clarke, my Group Leader, keeps his word when one of his SAs is promoted to Senior SA a year later and I'm put up for the vacancy. I'm on my way to London to be interviewed for it. It's a bit of a wasted journey really, my age and background education are more than enough to get me through. Even so, I get a good grilling on the experimental programmes I'm involved in. A week passes, then Alan taps me on the back and gives me the good news. He says a confirmation letter is in the post. The pay scale is £240–£490 per annum and I am on £360, about £6/18/- a week. I'm paid monthly and collect my wages envelope from 329, chuffed to find my pay packet now contains fivers, rather than just quids and tenners.

Paul hasn't been idle. As soon as the house purchase is complete she contacts Reading University Housing Department to register us as a boarding house. She leaves Trish with Mops while she spends time in town to sort out the cost of furnishing number 23. We come to the conclusion we will have to resort to hire purchase to get the place up and running by the New Year. Furniture and furnishings are still rationed but newly-weds and those setting up first homes are exempt. Most of what we need will come from two of Reading's major stores, Heelas and McIlroy. We shop around for everything else, from a scrubbing brush to full sets of china and cutlery, from a frying pan to bed linen. We are up to our ears in the never-never.

Mops agrees to look after Trish while Paul takes a job with Simonds Brewery in St Mary's Butts, just off Broad Street. They have become concerned about microbial contamination in their storage tanks and employ a microbiologist who needs a part-time assistant. Even though we take on students, she keeps this job going until our second child is due.

During what is left of 1951 we set about cleaning and redecorating our home from attic to outhouse. I try my hand at wallpaper hanging and wonder why there are bubbles all over. On the way back from doing a bit of shopping I stop to have a word with a neighbour. I mention in passing the problem I'm having with the wallpaper. She comes in and puts me right. Hey ho, how am I supposed to know that after pasting a length you have to fold in the top and the bottom to the middle, fold again and leave to soak for two or three minutes?

I do a bit of plumbing and need the mains turned off. The stopcock is a good arm's length deep by the front gate. I don't have a long-shanked tap key so it's shirtsleeve up and on to my knees. Down goes my arm and up it jumps the biggest toad I've ever seen. I fall flat on my back with this monster hanging on to my sleeve. You know how sometimes something happens to put ten years on your life? Well, I swear this one put twenty years on me.

Christmas 1951 rushes by as we put the final touches to the house. Three eighteen-year-old male students arrive on the first weekend of the new year and are duly installed: one in a modified living room; two upstairs in bedrooms/studies. It is the first time away from their families for all three and they are made very welcome. In no time, Paul has them chatting while she gets the dinner. The meal is ready at six o'clock and we all sit at the long refectory table for a substantial two-course meal, with cheese and crackers to follow. Paul encourages the students to swap stories and engage in a variety of topics. As we clear the table, she says this will be the routine and that it'll be traditional fish on Fridays, a roast on Sundays and leftovers on Monday. You know, cold meat with bubble and squeak.

Bearing in mind Paul won't be twenty-two until May, is pregnant, has a baby to tend, three students to cater for and is doing a part-time job, this is an exhausting time. But she bubbles away and in the evenings we find time to cuddle-up on the sofa and listen to the wireless, play records and chat about our day. My stint at Harwell finishes when I get home about 5.30. Work starts again with feeding Trish and getting her ready for bed, doing the washing-up and tackling a myriad odd jobs – from sundry repairs to doing housework and ironing.

It's a Saturday and I've been thinking – yes it does happen. Dad taught me to drive when I was seventeen and I obtained a provisional driving licence. Driving tests were stopped during the war and there was such a backlog of drivers waiting to be tested in 1947 that the government announced that anyone who'd held a provisional licence for at least one year since the start of the war would be given a full licence.

Our next-door neighbour Les is a self-employed taxi driver. We've got to know him and his wife Marion. I tell him of our money problems and ask if he can use an extra driver. I learn he has a private hire saloon, for which he has bookings way into the new year. This leaves his taxi idle at weekends so, as long as I agree to provide a service for his regular customers, pay for the petrol and clean the cab, he'll let me keep the takings. It's all a bit outside the rules but I know Reading as well as most and a street map will cover the rest. A lot of the weekend evening trade is with the military at Greenham Common and I have a feeling this will turn out to be a good earner. Sure enough, the taxi is a life saver, since we are only just covering the mortgage interest, the hire purchase repayments and the usual overheads, and Paul had stopped working at Simonds before May was out.

Money was so tight that we truly did feel down the back of the sofa cushions in the hope we would find the odd coin or two. Finding a florin, a 2/- piece, would have got both of us into the Granby cinema at Cemetery

Junction. I did so well with the taxi at Easter we did indeed go to see An American In Paris.

At 1.30 pm on 22 June 1952 I become the father of a son. Stephen Ronald Sambell, another curly head, comes into this world in sun-filled surroundings. Shift buses leave Harwell at 2.15 pm and I get on the one going back to Reading. I'd arranged for maternity leave and, with my fourteen days annual leave, I shall take over all home duties. It's a straightforward birth at home and Mops is keeping an eye on Trish until I get back. The students go down at the end of June, which gives me a chance to spend some time with Paul and the children. There aren't any postnatal complications and Paul rapidly gets back into the swing of things.

It is two years on, September 1954; two of our original boarders have gone into halls of residence so there are two new faces, new recruits as I call them. I drive the taxi as usual at weekends and Paul has persuaded Mops to look after both children so that she can do a couple of hours later in the mornings at Simonds. Christmas is upon us and we splash out a bit and treat ourselves to some goodies. Mops invites us round for Christmas Day dinner since she has boarders staying over anyway.

As an SA I am assigned to one particular Experimental Officer and take on the day to day management of a project relating to the effects of gamma radiation on the creep resistance of nickel alloys at temperatures up to 1,000 Centigrade. The results are reported with a secret grading every six months and my immediate boss includes my name as co-author. At the end of three or so years of support on this project Alan recommends me for promotion to SSA, which I duly attain in 1957. My pay increases to £725 per annum and we suddenly find ourselves with more than enough to make ends meet.

Early on in 1956 Alan tells me he has an outstation laboratory and he wants me to provide some support there a couple of days a week. Where is it? Only just across the road from home, at 18 College Road. You could knock me down with a feather. It seems Dr Karl Mendel, a chemical engineer refugee from Germany, had applied to the Ministry of Supply in 1948 for a job, on Alan's recommendation. He'd been accepted, with the provisos that a facility would be created off-site and that he would report to Alan. I cannot believe my luck. On two days a week I'll get up a bit later. This doesn't happen, of course. We get to know Karl by asking him over at weekends. It's a bit of an ordeal since his English still isn't very good and his accent is very pronounced. There's a quietude in his manner, a sadness in his bushy-browed grey eyes; he lost his family in the Holocaust but survived his imprisonment in Terezin, a concentration camp in Czechoslovakia, although it left him with a limp from the ill-treatment.

Terezin is a former military fortress comprising The Citadel and an adjacent walled garrison in what is now the Czech Republic.

Terezin

No room has ever been as silent as the room
Where hundreds of violins are hung in unison

This two-line poem by Michael Longley is well worth analysing. I challenge you to have a go. You will find my own reactions to it at the end of this book.

Karl tells me that two Polish migrants are living with him for a few weeks and, with winter setting in, he wonders whether we have any suitable clothing to spare. I promise to look through my wardrobe. As I get ready for bed I do just that and find a spare overcoat and two thick jerseys, a pair of gloves, a couple of scarves and pairs of woolly socks. I'll take them over tomorrow, after I've been on the scrounge. Before they leave, the two Poles come across to say goodbye and to thank us for the gifts. Neither has much English but the shadows in their eyes speak for them.

Alan Blainey migrated to South Africa in 1958 and the outstation was shut down. Karl returned to West Germany where he still had relatives.

We have a postgraduate boarder, Dave, a tall gangly man with a mop of blond hair. He's quite a wag and provides a bit of uplift at the dinner table. Zeb is staying with us for a few days while he recovers from surgery. His lower jawbones had been adjusted and wired-up to keep them in place while they bond. Mum cautions us not to make him laugh, since it may disturb the wiring. Dave has a job staying in line, which only makes it more difficult for us all to keep a straight face. To ease things Paul suggests we see a film. *The Vikings* seems a safe option and she arranges for a neighbour to babysit. The lodgers are invited to come along and Dave takes up the offer.

As we walk back home along Minster Street, Dave starts leaping up and down, bawling, 'Odin, the Vikings are coming.' Next thing, we're chasing down the road yelling our heads off, brandishing imaginary axes. Suddenly, Zeb stops in his tracks and puts his hands to his face. No popcorn for guessing what has happened. It is back to the hospital for him – and a serious telling-off for me from Mum and Dad for being an irresponsible twenty-seven-year-old.

'You really should have known better.'

Another tick in the black book.

Curiously enough, the 'Odin' war cry has stood the test of time. Zeb and I have never been close and have met up just a few times since then. The yell seems to act as a catalyst; one of us will get it in first and the intervening years are stripped away, giving me the feeling he's just come back from the shop down the road.

Trish is now seven and has been at Alfred Sutton Infants since autumn 1956 and Steve has just started there. Our third offspring, Richard Andrew, birth date 28 July 1956, is two. He is the heavyweight of our tribe, ten pounds four ounces at birth. Another curly head, and I do mean tightly curled.

Tricia and Steve are turning out to be real devils at home. They put Richard in his walking harness and string him up in the pear tree at the bottom of the garden. They then call us out saying that Richard needs help, we rush out and I nearly have a heart attack when I see what they've done. At first glance it looks as if he'd hung himself, all three have a good laugh when they see we appreciate the joke.

They decided to descend from the landing window on to the glazed roof of the front porch, using a length of cord loosely tied to a newel post, with no thought as to what might happen and what to do next. This time though we don't see the funny side; Alf lying there with his head all swollen has just flashed before my eyes.

Stephen's nose looks a bit swollen on one side and his breathing becomes noisy. It doesn't improve so we decide to take him to the doctor. Our consternation is quickly followed by relief and embarrassment when the doctor flourishes a pair of forceps under our noses, holding a collar stud. Steve denies all knowledge of course.

Mind you, it isn't all one-sided. There's another twine scenario the day Paul wants Richard's cot to go in the garden as a playpen. In her wisdom she decides to lower it down out of the window on a length of cord. It snaps, of course, and the cot falls on Tricia's head. It's off to casualty at Battle Hospital in double quick time, where a gash on her forehead requires stitches. She still has a small lump as a legacy.

Soon after starting school, Trish becomes friends with Patricia Hill who lives opposite us. Their birthdays are only two weeks apart. Then I learn that her father, John, also works at Harwell and the reason I hadn't met him before was because he drives to work. He is Senior EO in the Atomic Particle Separation Division, and works irregular hours, so a car is essential. I pop across to say hello and after swapping bits of our backgrounds it transpires that John is Scout Master of the 73rd Reading Scout Troop, based at the Park Congregational Church on the corner of Palmer Park Avenue. When he learns I've been a Scout he asks me to join him as an Assistant Scout Master. It is a big troop and he does get a lot of backing from church elders and has an assistant, but the latter hasn't much scouting history and John is sorely in need of some experienced support. I collapse on the settee when I get back and talk to Paul about John's offer and get her enthusiastic go-ahead. Over the next five years John becomes a great friend. He served as flight engineer and gunner in a Beaufort bomber during the war and has both feet very much on the ground. His is a serene outlook on life, with a smile never far from his chubby face and he's very practical and conscientious in his endeavours. He is married to Nan, whom he met while serving in the RAF; Nan was in the Women's Auxiliary Air Force as a plotter and her personal forte is an ever-smiling, bright-eyed face.

Summer camps become our fortnight summer holidays. Various incidents stay with me, such as: everything being stowed in Catt's Removals pantechnicon with the troop piled in on top; the aerial rope slide where Paul had to have a go and thumped the ground so hard she had a tender backside for a couple of days; the tree catwalk, about fifteen feet up, and one of the younger Scouts being so petrified we had to get him down in a harness; John's Vauxhall giving up the ghost brake-wise and trundling down the slope, right through our cooking fire and evening meal and ending up in the ditch – we had to get the farmer to pull it out with his tractor; Skip Ireson, our Group Leader, introducing us to his cheese dreams during final evenings around the campfire; the sketches each patrol put on during these occasions; Paul leading the singing of the campfire songs and showing each patrol in turn how to make a dough twist that would stay on a long stick and cook over the camp fire – a bit charred but scoffed down as if it was birthday cake; the shelving pebble beach at Beer, where we adults had to dog-paddle and form a cordon within which the scouts could 'safely' enter the sea.

In Easter 1963 some of the troop took the bold step, in four cars, of spending a fortnight in Kandersteg, home to the International Scout Skiing Hostel in Switzerland. I put on a pair of skis and learn how to cross-country in a matter of days. The most difficult exercise was discovering how to stop in time.

Family holidays are fitted in, with stays on Hayling Island in a caravan, at Sandown on the Isle of Wight and Rhossili on the Gower peninsular in South Wales. It is at Rhossili that Steve manages to break his toe on the only rock sticking up out of a couple of miles of pristine sandy beach; and I manage to break a Thermos flask, buy another, and then smash that one as I negotiate the sand dunes behind the beach. Philip, number four in the hierarchy of my children, came on one of the holidays at Sandown with his leg in plaster, having managed to get it jammed under the roundabout in the park playground two days before we pushed off. What a to-do keeping the sand out of it.

After about two years with the scouts, Paul is approached by the church wardens to form a youth club, subject to some training. She hums and haws and I know why. She is reluctant to become formally involved with the Church. We talk about it at great length and finally thrash out a solution, which she puts to the elders. Providing she can operate through a committee of youth club members, with the chair reporting to the wardens then she is prepared to give it a go. This is okay with them and within three months the club attracts about forty members.

The Park Youth Club Committee meets at our place every couple of weeks. Football, hockey, netball and table tennis teams have been registered with the relevant Amateur Associations. A stall will be set up at the autumn bazaar to help raise funds for the upkeep of the church premises. A panto is planned for Christmas 1960.

The club thrives and in summer 1962 club members and our family settle in on the Mary Watson Barge, a Youth Club Association floating holiday venue on the Thames at Abingdon. Slung hammocks and deck scouring are novel features, which some baulk at but it all goes off smoothly, almost. Despite warnings not to go barefoot on the deck, harum-scarum Jane Parr runs a sizeable splinter into the sole of her foot and is taken to casualty; followed by Tricia and Steve deciding to chance their arm in the dinghy moored to the barge. They are in midstream as a pleasure steamer appears round the bend and pandemonium sets in. There is much shouting of advice that I'm sure isn't registering. They row and steer on the same side and start to go round in a circle and the captain sounds his horn. The dinghy looks awfully small, fortunately the steamer has only just left the lock and is barely under way, the captain goes into reverse and the would-be adventurers sort themselves out. This is a bit breath-holding for a while; it's my turn to tick the black book.

The Christmas show in 1964 is really quite professional, a mixed panto and variety affair. The club has its own group of four musicians, two electric guitars, a bass and drums, who bring the house down. They only just missed out on winning a regional youth club competition – points were lost because the drums were set too far back in the group's spatial arrangement.

I've become reasonably proficient on the classical guitar, having bought one from a Harwell staff member who threw in lessons. Paul and I form a creditable duo and her renderings of 'Summer Time' and 'Moon River' go down very well; as does Julie Timms' solo during the interval, of the popular 'Come Outside', recently released by Mike Sarne and Wendy Richards.

In autumn, a night hike and treasure hunt are organised in Mapledurham estate, with a sleepover at our place. About fifteen clubbers are given supper and bedded down in spare rooms and passages. It is great fun, despite the fatigue, and is duly in the next report to the church wardens. They come to the conclusion that Paul and I were wrong to allow non-segregated sleeping arrangements, even though it was open plan, with due vigilance on our part. Paul is immediately up in arms and we ask the Reverend Clamp to oversee a meeting between us and the church committee. This meeting doesn't go in our favour and Paul's resignation is accepted. I feel it is right that I leave the Scout Group since I'd helped organise this outing. Six or seven months pass. There's a new youth club leader but the members do not ride this trauma and the club folds.

In 1961, I'm upgraded to Experimental Officer (EO) as a result of two occurrences. First, after attending Reading Technical College on day release in 1958, I now have a Higher National Certificate in Applied Physics; following a reorganisation of the Metallurgy Division, the bulk of the workforce moves to a new building, 393. Our division head is Monty Finniston. Freddie Clarke is still our Group Leader, a large group of six sections specialising in non-metals. The spectrum of materials ranges from concrete, to single crystals, to composites such as carbon fibre-reinforced glass and ceramics, as distinct from reinforced plastics. The second occurrence is being assigned to a new Section Leader, Dr Jack Williams, a Senior Scientific Officer and a fundamental scientist researching the crystallography of radioactive oxides, nitrides and carbides. I report directly to Jack and my initial research is on the crystallography of the uranium/carbon system, during which I establish that uranium dicarbide exists. I have my own SA for this project. It is the publication of this research with me as lead author that does the trick.

I also develop a vacuum hot-pressing process and demonstrate that magnesium oxide powder, which is opaque when hot-pressed in air, can be made optically transparent if the densification is carried out in a vacuum. In the same period I demonstrate the ductility of single crystal bars of magnesium oxide that have been chemically polished and so are free of surface defects. I film the bending of chemically polished, single crystal bars through ninety degrees without fracturing.

Freddie is a born entrepreneur and designs a library of topics connected to the Group's work. After a bit of arm-twisting, muggins here 'volunteers' to keep the library updated. This entails scanning relevant journals and proceedings for items relating to our research projects. It is time-consuming and has to be done in my own hours. The Group meets once a month so that I can recommend reports and papers of interest and pass on digests of information. It's a real pain but I develop an in-depth overview of the Group's programme. The publication of my work on irradiated magnesium oxide in the *Philosophical Magazine* results in promotion to Senior EO.

In 1973, yet another new building is completed and Materials Development Division (MDD) 552 is created. By now, UKAEA has devolved into governmental and commercial activities in order to become more self-supporting. Within MDD, Freddie Clarke is made head of what is called the Ceramic Centre and I join it as a Group Leader, with a remit to obtain commercial contracts.

We come to the notice of industry in general when a *New Scientist* editorial on the Ceramics Centre is published. My work on producing optically transparent magnesium oxide and carbon fibre-reinforced ceramics had been successfully patented and was included in this review. By 1978 I am responsible for generating over one million pounds per annum of non-government research, with companies as diverse as ICI, Thermal Syndicate and Rolls-Royce. All in addition to my work on the immobilisation of low- and medium-level radioactive wastes. This results in a cross-over promotion interview for Principal SO and is where my muggins stint proves invaluable. The interview board are as much interested in my broad knowledge of the Centre's programmes and its prospects, as in my own contribution. I set a precedent by being the first non-graduate PSO in the Authority.

In 1977, £1 was equivalent to over £5 in 2021, so my contribution to the commercial portfolio of the Ceramics Centre was no mean feat. On top of this were my patents for transparent ceramics and spun ceramic fibres, both of which were accepted in the US, a more difficult thing to do then. Patent and financial returns from commercial licensing belong to UKAEA of course, but as the principal patentee I still have one of my mandatory $1 registration fees in the form of a 7/- postal order. Appendix 2 lists the patents I was involved in.

In 1979 I head up European Research on the Disposal of Radioactive Wastes (DRW) and I report directly to the UKAEA Directorate and the Commission of the European Communities. Freddie, effectively Division Head of the Ceramics Centre, and Jack, now a Senior PSO, move on to the Directorate at Harwell and Dr David Livey takes over from Freddie. I no longer identify with a group and report directly to David on my own projects on radioactive waste disposal and contracts for commercial projects. In order to work, travel and entertain at the same level as my European colleagues I am given the expenditure level of a Senior PSO.

Dr Roger Davidge, a PSO, has taken over Jack's group. Roger is a classic scholar-scientist, tall, black-haired and with a full beard. He has a reserved nature and is not much given to fun and games. His classical approach to the fundamental behaviour of ceramics under stressful conditions is being recognised worldwide. We team up to complement each other on the engineering properties of structural ceramics and the economic/conservation advantages of using them. I present a paper on this work at a Harwell conference on 'Conservation of Materials'.

This excursion into conservation prompts me to explore sulphur as a structural material rather than as a chemical source. Information from published reports and proceedings leads to a presentation to the Directorate, endorsed by David Livey, recommending a research programme which, after a bit of tweaking is used as a blueprint for a full-blown project. I am expected to head up this programme but the paper is my final action before requesting early retirement.

In the meantime, other things have happened. John and Nan Hill moved to a big detached house on Wokingham Road in the early 1950s. It wasn't long before I found myself helping him to dig out the foundations for a swimming pool. Les, Marion and children emigrate to Australia in the mid 1960s and I no longer drive the taxi.

I hear of a vacancy for a weekend barman at The Pheasant in Winnersh. I try my luck at it and get taken on. In 1966 my boss, with his eye for the main chance, acquires five tickets for the Football World Cup Final at Wembley, the finalists being England and West Germany. I have no abiding interest in professional football but the offer of two tickets for a fiver a go is not to be sniffed at, even though it is most of a week's wages. I ask a colleague at Harwell, who is a great enthusiast, if he wants the other one. He bites my hand off.

You have to personally experience the atmosphere on such occasions. The passion is remarkable and it does get to you. Well, it got to me. Two all at full time and then England score their third goal in extra time. I go bonkers with the rest. Then the fourth goal is added and the world goes mad. My companion grabs hold of me and we do a jig on the spot.

I have quite a collection of memorabilia of that event, including my ticket, programmes, photographs and the autographs of the whole team and the manager. 'Some people are on the pitch ... they think it's all over ... it is now!' were the immortal words the BBC commentator Ken Wolstenholme exclaimed as England's Geoff Hurst became the only man to score a hat-trick in a FIFA World Cup Final.

Philip Roger Sambell arrives on 2 December 1959, straight-haired for a change. Apart from the leg-in-plaster bit nothing much happens to him, just the humdrum of family life and being bossed about by his sister.

By the time he starts at Alfred Sutton Primary School, Phil has a good sense of humour and later shows he's inherited Paul's singing voice and her ability to turn a conversation on its head. He is a good academic and moves on to Reading School. When we move closer to Harwell in 1972 he attends King Alfred School, a grammar in Wantage. From there he gains entry to Hull University to study economics. After several years in the entertainments industry in London, as a designer of and advisor to pubs and nightclubs, he toys with the idea of having his own pub.

We are now a sizeable family of six and my promotion to SSA in 1957 means we can think about not taking on students so that the children can have their own rooms. Weekends develop a pattern of togetherness which we all look forward to, particularly when we buy our first television set in order to watch the 1964 Tokyo Olympics.

That first TV was in black and white. Colour television did not take off until 1967 on BBC2, the first channel in Europe to transmit colour regularly. We changed to colour in 1972 after our move from College Road.

Saturday afternoons are tea-trolley, pastries, hiding behind the settee when the Daleks rampage, followed by dinner. Weekday telly for the children is well catered for with *Watch with Mother*, *Jackanory*, *Magic Roundabout*, *Blue Peter* and many more. Paul and I settle down to watch the Wednesday play and I thoroughly enjoyed two in particular, 'Up The Junction' and 'Cathy Come Home'. We also follow series such as *Emergency – Ward 10*, *The Likely Lads* and *Dixon of Dock Green*. All interrupted by further additions to the family.

Lindsay Anne comes into this world on 21 May 1963. I sit on the stairs with the children, Trish comforts Phil as Paul lets rip and we wait with ants in our pants for the reappearance of the midwife and the welcome news that all is well and we can go in. Hot on Linds' heels on 16 July 1964, Robert Alfred arrives. Paul has no postnatal difficulties and picks up where she left off. She is supposed to have been on contraceptive pills for the last couple of years but is very lax about taking them, with the observed results. The growing number of my offspring is a standing joke among my friends at Harwell. I'm going on thirty-six and I think six is quite enough. There's nothing for it, I shall have the snip.

Tricia, at thirteen, becomes a real stalwart and gives Paul as much help as she can. She flourished at Southlands Girls' School and became a competent piano player with a solid technical understanding of music. Sports-wise she was the netball captain. She excelled in her general subjects and entered Rolle College Exmouth, a Plymouth University Campus, where she obtained a Bachelor of Education Honours Degree in 1976, despite having to take a year out to undergo a spinal operation. Trish joined the Guide Movement as a Brownie and went on to Guides, where she tried to chop her leg off when collecting firewood. Later in life she became a Cub Scout Leader and progressed to Devon County Commissioner for Beavers. She has taken up hand-bell ringing as a serious hobby, giving her the opportunity to travel far and wide.

In 1967 we get our first car, a Vauxhall Victor Estate. I dislike the colour, either that or I'm having another Moses fantasy, because I'm changing it to British Racing Green! By this time, Mum and Dad have retired and are living in a newly built, four-bedroom detached house in Decoy Drive, Eastbourne. We drive there several times, loaded with children and beach paraphernalia.

Mum and Dad come to look after the children while Paul and I go on our first holiday abroad, to Alassio on the Italian (Ponento) Riviera in Liguria, courtesy of Martin Rooks Holidays. Wet behind the ears doesn't really describe our lack of experience regarding competing with the Germans and French for meals and beach space, but we soon cotton on. We sightsee and spend a lot of time on the beach. Swimming in lukewarm water is a nice change.

On the day before we are due to leave I decide to take a longer swim, only far enough out to avoid the lilos and the splashing. Then I make for the pier, the Pontile Bestoso, about a quarter of a mile further along the beach. I swim steadily but slowly, take a rather long breather at the jetty and return as I came. As I wade ashore I see people gathered around where we are sitting. My immediate concern is for Paul, but before I can enquire she jumps up and throws herself at me. Safe in my arms she lets out an almighty scream and bangs me on the chest.

'Don't you ever do that to me again,' she bawls in my ear as the onlookers clap and cheer.

As I referred to earlier, this was the second time where my lack of consideration caused tremendous worry for a loved one. Paul really thought I'd drowned, being so long away.

A Red Letter Day. On 15 November 1971 our Britannic policy matures. Twenty years have sped by and the value of the house has increased seven-fold. Things have changed a bit in our neighbourhood. Familiar faces have passed away or moved elsewhere. Our long-time friends John and Nan Hill now live in East Hendred, near Wantage. John is retired, having spent his last years at the Culham Fusion Reactor Research Laboratory close to Abingdon. I think he looked forward to retirement and tending his extensive 'olde worlde' cottage and garden and the very old church clock which has no face.

John's commitment to the upkeep of the faceless clock is amazing. It has to be wound up every day for a start. The clock was made by John Seymour of Wantage in 1525. The faceless clock not only strikes every quarter-hour but it also plays 'Angels Hymn' by Orlando Gibbons every three hours.

I worked with John and his team when the Ceramics Centre was employed to study the stability of a range of ceramics and composites at the high temperatures involved in fusion reactors. My briefs were to assess carbon fibre-reinforced aluminium oxide as a liner and the handling of the gaseous radioactive waste.

Conventional atomic energy results from splitting atoms into smaller particles. Many more times energy is produced by fusion rather than fission. A plasma of positively charged nuclei and free electrons, as in the sun, needs to be formed – the so-called fourth state of matter. Fusion occurs when pairs of nuclei are forced together in the sun's magnetic field. Two particles make a bigger one, a helium ion, with a release of heat, 15 million–100 million Centigrade at the core of the sun. This is due to the density of the sun's core being many times greater than that of gold, even though it is a plasma. It requires high temperatures and super-efficient electromagnets to do this artificially. But if all this can be sustained, the net result is a very high gain in heat energy; with the added bonuses that there is a limitless supply of water from which the fuel, deuterium and tritium, is separated and that the radioactivity of the waste product, helium, is very short-lived.

I saw a photo of a monolith that resembled a gigantic ring doughnut stood upright. The term 'toroid' came to mind. I have attempted to create a free-form poem describing in non-technical phrases, the process of induced nuclear fusion. I chose 'Tokamak' as the title since those readers not recognising this acronym might be intrigued enough to read on.

***Tokamak*†**

Consider a toroid, a hollow O-ring if you will
Consider then two captive atoms
A pair among zillions of whirling dervishes
Super-magnets strip off electrons
A plasma is created
Our two positive nuclei now mutually repulsed
Are force-fused together in a magnetic pinch
The energy released is very small
But fusion of the zillions is taking place
And useful heat is won

†Tokamak: An acronym of the Russian for a toroidal magnetic fusion power reactor, **TO**roidal'naya **KA**mera s **MA**gnitnymi **K**atushkami.

We decide to move closer to AERE and start house hunting. Eventually, we come across a new detached house in the process of receiving the finishing touches. It is in The Croft, West Hanney, and sits in a third of an acre of field, with elm trees and a footpath along the far side.

East and West Hanney are midway between Wantage and Abingdon in the Vale of White Horse. The name dates back about 5,000 years. The Vale was originally marshy and the 'ey' ending of the toponym is Old English for 'island', hanena-ey being 'island of wild turkeys'. Other examples nearby are Goosey, Childrey and Charney Basset. Flooding occurred in 1974 and canoes paddled to and fro. It is just as well the pavements through the Hanneys are raised.

We move in mid-1972 and the house is christened 'Song of Summer', reflecting Paul's love of Delius and our shared fondness for music and classics in general. It has three main reception rooms, a small study, a toilet, a large kitchen and a utility room. The garage is attached, with a bedroom above. There are three more bedrooms, a box room, a bathroom and a loft. I can look forward to a lot of decorating.

The ground behind the house is more like a builder's yard, an urgent clearance project for me. I proceed to create a patio outside the sliding doors in the living room, a fish pond beyond that with a rockery surround, turf for a lawn, edged by a low wall with a step up to vegetable plots and flower borders, mini-orchard and soft fruits – all told, a couple of years' work.

I think, 'Ah I can sit back now.' Not on your life. Four things occur: 1. Dutch elm disease strikes and I have ten dead elm trees to deal with; 2. We decide to get a couple of cats and a puppy; 3. A bit of 'what goes around comes around' surprise, I'm helping to dig out the hole for a swimming pool in the grounds of the village school – how this happens is a bit of a mystery to me but I think Paul is behind it; and 4. The wheel goes full circle when a parent discovers I have Scouting experience – Paul again I suspect.

A Scout group has just been resurrected in West Hanney and the leaders are desperately short of experience. Please will I take on the role of Group Leader. How can I not say 'Yes'? I tell the leaders my reason for resigning from the 73rd in Reading. Being some fifteen years ago this is laughed out of court, but I insist they make the District Commissioner aware of it. This is to no avail and before I know it I'm newly kitted out with beret and long trousers – Scout hat and shorts are long gone, as are staves – and holding my first leaders' meeting at the house of Scout Leader Angela Cousins. Her husband, Lionel, a quiet considerate man and chairman of the parent's committee, is also present. I take an Oxfordshire Scout Headquarters course for Group Scout Leaders. Within no time Paul is invited to become helper-in-charge of equipment, catering and mothering. Robert and Adam (Chad), the younger son of Angela, are both in the troop and buddies. It is a proud day for us when they are presented with their Chief Scout's Award badges by the Oxfordshire County Commissioner. Angela goes on to become a County Commissioner and gains the top Scouting accolade of Silver Fox for her contribution to the Movement. I have to pull out in 1979, due to my workload and frequent absences from home.

We have some great fun parties at the Song and three parlour games in particular are my favourites.

Escape the Blanket, where you are blindfolded, made to lie on the floor and covered with a large blanket, then you have to find your way out. Easy peasy? Not if there is someone at each corner ready to shift the blanket if you get too close to the edge. One or two victims panic a bit.

Jump, again you can't see and are made to stand on a plank supported by a book at each end. Pretence is made to place another book under each end of the plank while two others support you, after several such pretences you are asked to stand up straight at which moment a book is held fractionally above your head and of course as soon as you touch it you duck, thinking it is the ceiling. Then you are ordered to jump, most don't have the nerve, despite all sorts of cajoling. Imagine the consternation when the blindfold is taken off to find you are only three inches off the floor.

Charades is always a good laugh. Paul comes into the room, sitting astride Lionel's back as he crawls in on hands and knees, his head waggling side to side and his tongue hanging out. None of us twig the charade is 'verging on the ridiculous'. Get it?

When we make the move to The Croft Richard still has a year to do at Reading School, so rather than uproot him at this critical point in his education we arrange for him to become a boarder. After his exams he joins us in West Hanney and starts a stop-gap job in the flour mills in Wantage. In 1973 he follows me to Harwell as an Assistant Scientific Officer in Chemistry Division.

We think long and hard about schooling for Lindsay and Robert. They are still primary school age and there is a village school in East Hanney. Bearing in mind my own experience of such schools, we make exhaustive inquiries about the quality of education there. Because of the influx of professionals with educational aspirations for their children there has been a lot of pressure over the years to upgrade the curricula of village schools in the Vale and we become confident our two will not be held back. They both go on to King Alfreds, in Philip's footsteps, and do well enough to give them a good start on leaving; Linds goes into banking with Lloyds in Wantage, and Robert into superconducting magnets for research with Oxford Instruments.

Archatso Serenity is the name on the puppy's pedigree, born 21 March 1981. To us, she is simply Lara, one of the gentlest of medium-sized breeds, snow white with a smiling face. When she is about a year old I enrol for dog training sessions, including dog handling when entering competition show rings. Although the shows are very competitive, the meetings are friendly and, at the outset, I receive lots of advice and tips. Over the next five years we go on to win best of breed and best dog in show several times at local and Oxfordshire/ Berkshire Kennel Club events – Crufts was never on the agenda, it's for professional dog breeders.

The Samoyed is an ancient breed, originating in the Samoyedic nomad tribe in Siberia and is used for herding reindeer and hauling sledges. They are family dogs and love a cuddle. Their happy features earn them the nickname 'Smiling Sammies'. They can be wilful and it requires patience to train them, but eventually it sticks and they 'show' very naturally. Samoyeds are reputed to be noisy, but the only times Lara barked was to give an alert. The snag was she then approached 'intruders' with a wagging tail and a lolling tongue. Paul and I were devoted to her. She lived to the ripe old age of fifteen.

A favourite walk for Paul and me is from the Uffington White Horse to Wayland's Smithy, a long barrow on the Ridgeway. You can see for miles across the Vale. We'd take Lara and she'd chase the rabbits, who always managed to reach the safety of their burrows in the hedgerow a split second before she made it. Sadly, myxomatosis took its toll of the Ridgeway rabbits and this source of excitement and exercise was lost to Lara. The Ridgeway is noted for its horse riding and the awesome sight of up to twenty racehorses thundering by as they stretch their legs is something to remember.

The Uffington White Horse is a minimalist outline dating back to the Late Bronze Age (1000–700 BC). Metal working was believed to be magical and legend has it that Wayland's Smithy is the home of Wayland the Saxon, an elfin god of metal working. The story says, if you tether your horse at the long barrow and leave a few coins you will find it freshly shod the following day.

In November 1972 Lindsay takes a phone call from Peggy, Mop's daughter. She tells us that Peggy's brother Ron, a chemical engineer had died from a massive heart attack while working for his firm in Italy. When Lindsay gives us this sad news I feel like crying. He was a big man, my best man, a bit ham-fisted when it came to DIY, but oh so gentle and caring and with a good sense of humour. His wife Richenda, Genda to the family, and their son Colin, invite us to the funeral at Swanton Morley, a sprawling village about seventeen miles north east of Norwich. It is Genda's home ground and where she now lives. Peggy comes with us. It is a drizzly day, a day for mourning.

Because it has travelled, the coffin is lead-lined and very heavy and hence requires eight pallbearers. Planks have been laid down either side and across the grave and the four webbing straps for lowering the coffin are already in place. As we gather round, the bearers lower the coffin onto the cross planks, pick up the straps and take the weight of the coffin off the cross planks, which are then withdrawn. The Minister has hardly begun the requiem when the bearers on one side stagger as the wall of the grave gives way and the plank they are standing on starts to slip into the hole. They desperately hold on but the weight is too much and they follow the coffin into the grave. This causes the four on the other side to join them as the grave wall on their side disappears. Two of the first four are now trapped under the coffin, which is canted with the head-end stuck in the end wall of the grave. No one has the faintest idea how to cope, but I do ask the Minister to at least stop incanting. Some of the men present galvanise into action and get the second four out. It then requires much more care to assist the first four because of the danger the coffin will slide down the end wall and crush the two underneath. Eventually all are got out safely, the only mishap being a pulled muscle. The coffin is re-aligned and the service completed.

Back at Genda's house there is meant to be the usual chatty wake as family members catch up with each other, but everyone is still shaken by what's happened and the mood is solemn. Peggy and Genda in particular look distraught, so I suggest to Peg that perhaps we should be on our way. She is sitting in the back of the car with Paul hugging her, when I hear her start to chuckle. Thinking she might be a bit hysterical I pull over and turn round. Peg lets out a good laugh.

'Everything alright, Peg?' I ask.

She nods her head, giggles and wipes her eyes, 'You've got to hand it to him. Awkward to the last.'

Which sets all three of us off.

By latching leave on to some of my many official overseas visits from 1978 to 1985, Paul is able to accompany me. This is a most rewarding period for both of us and there are some notable incidents.

Our first visit to the US in 1979 is primarily for me to attend a conference being held by the American Ceramic Society in Cincinnati. Many of the delegates have their wives along and in no time at all we are invited to social gatherings in the evenings, during which we gravitate to two couples in particular. They are opposite to each other in backgrounds but we seem to provide a common link and the two couples become our lifelong friends and stay with us several times over the years. One of the suppers was buffalo steaks, cooked rare of course, but that didn't stop Paul from objecting to eating raw meat.

We go on to New York, where I have discussions at the National Bureau of Standards in Manhattan with regard to the rationalisation of techniques for radiation measurement. We are both shocked by the murder of John Lennon whilst there. During this visit Paul takes a wander down Broadway and goes into a relatively unsafe area, a patrol car pulls up and a policeman questions her need to be there. After a bit of a chat they decide to give her a ride back to our hotel. There you go.

We take this opportunity to fly on to Pittsburgh, to meet up with Paul's nephew, Peggy's son, Clifford and his wife on our way to Chalk River Nuclear Laboratories, where I have discussions about high-level radioactive waste immobilisation.

In 1982 we go to Cadarache, Europe's largest atomic energy research establishment, some forty miles north west of Marseilles. This is my third visit to present a progress report on Radioactive Waste Management. Paul is considered a notable guest and we are given the Tower Suite. It was used by Charles de Gaulle during his presidency, hence the unduly high, double security doors – he was nearly seven feet tall in his habitual military hat. After our first night in the bedroom we have to confess that the bed had collapsed. There is much side-of-the-nose tapping, 'formidables' and 'ooh lah lahs' at the breakfast table. Paul is right there with them.

'It's the French air that's to blame, we just went along with it,' she retorts in very passable French, hands on hips and hair tossed. My French colleagues love it.

In Rome we get involved in an altercation between two taxi drivers who had both turned up at our hotel to take us to dinner with my Italian counterparts. Paul's smattering of the language confuses the situation and almost results in us being dragged off separately. She gets over this by asking the receptionist to order another one. I resort to giving the two irate taxi drivers generous tips.

While in Belgium, during our own time we visit the shellfish restaurant area in Brussels, one street in particular is noted for mussels. We both enjoy these but not the succession of begging-bowl troubadours, who get partway through a tune and stand there at your table waiting for a gratuity before moving on to the next eatery. This gets right up Paul's nose and after three or four visits by these 'entertainers' she holds on to the jacket of a guitar player and sings to the melody he is strumming. She gets a loud round of applause, a peck on the cheek from the guitarist and a call for an encore.

Our second visit to the States, in 1984, takes us to Washington DC and the opportunity to visit the White House – no incidents fortunately. I also have to go to Seattle, where we lunch at the Copacabana in Pike Place Market and visit The Waterfront. We have dinner on the third evening in the revolving restaurant at the top of the Space Needle. A three-piece group is playing and after our meal we get up and have a bit of a dance, during which Paul goes over and asks them to accompany her singing 'All You Need Is Love'. Several other diners join in once they realise Paul is a very good vocalist and the requests keep her busy.

We attend another Cincinnati Conference in 1985 and two other Harwellians are attending. They've hired a car for the four of us to go 'Down South' to a meeting at the Savannah River Heavy Water Site, South Carolina. During an evening meal with site staff one of them, referring to our room, asks Paul what she thinks of the boob-tube. Her quick wit and a very straight face come into play, she draws herself up and suggests this is somewhat personal.

'Excuse me, I've no idea. I wear a bra.'

There is a moment of raised eyebrows before the penny drops.

On the occasion that Harwell hosts a meeting of DRW projects in 1984 we invite the members I know to dinner at our house. There are eight, including me, and Lindsay stays to help Paul prepare and serve the dinner. It makes me very proud to see Linds handling this situation very smoothly, to the extent that the group insists room is made so that she and Paul can join us for the meal. Thank goodness for the long refectory table and ten place settings from the student/offspring days.

Harwell's isolated location developed several facets: the original houses and messes built for officers and other ranks are now used by AERE staff; a large number of prefabs have been built for married couples; there is an off-site general store, a hairdresser and a pharmacy; on-site, there is a restaurant, a cafeteria and a goodies shop. Richard and Steve live in B-mess. When Stephen and Elizabeth marry they rent a prefab.

It is just as well these facilities exist. The winter of 1963 was very cold in the south and snow drifted so badly, the roads across the Berkshire Downs became impassable. I was stuck in AERE for two nights. The restaurant was kept open and dinner was supplied free. There were some almighty big snowball battles between the groups of Metallurgy Division. It all happened again in 1974.

Many staff who maintain the infrastructure and grounds come from surrounding villages, where horticulture is a required skill, hence there is a vigorous and competitive gardening club of some 300 members. Once our garden is under control I decide to join the club. It takes a couple of years of soil improvement and lots of spontaneous advice from enthusiasts to reach a stage where I could think about entering the annual show; a grand affair, with marquees, side shows and sundry classes to enter. My first attempts are rather pitiful, even for the novice classes, but by 1975 I'm able to grow winning entries of vegetables and flowers, with dahlias as my speciality. The most challenging class is to show a collection of five different vegetables, six of each. Uniformity and quality within a species is the objective: long straight runner beans; long tapered carrots and parsnips; globe-shaped cabbages; potatoes of similar size with unblemished skins. All pest free and all coming to perfection at the same time. I win this competition in 1979. My dahlias come up trumps two years running.

There are also football, cricket, snooker and arts clubs. I play cricket for Metallurgy Division and later for the Ceramics Centre. Richard joins the rugby and snooker clubs and is in a very successful team in the Didcot and District Snooker League.

Stephen is into dramatics and ropes in Paul and Richard to participate in a production of Dylan Thomas' *Under Milk Wood*. In 1980 the Ceramic Centre produces a variety concert, *Monday Night at Aston Tirrold*, a spoof on *Saturday Night at the Palladium*. Aston Tirrold? A small village beyond Blewbury, really in the sticks. We go there simply because the village hall is available. I compose an anthem for the occasion and actually persuade the Director of Harwell and his wife to sing along with us to the tune of 'When the Saints Go Marching In'.

It is hilarious. Paul's voice is now much richer and I am much more proficient on the classical guitar. We give a good account of ourselves.

Chapter 10

A Swan Song — Did This Really Happen?

Remember I said Philip had mentioned the desire to have his own pub? At fifty-seven I'm at the top of my career grade. I've had my fill of travelling, in particular to-ing and fro-ing to Brussels; and of being boss and editing, rather than doing my own original research. Paul and I talk this over and, out of the blue, agree we could help Phil achieve his objective. Will we have the capital to do it? Retiring age from the authority is sixty, so I'd be sacrificing three years of my retirement lump sum – no big deal. Our small mortgage has no foreclosure penalties and our house has at least quadrupled in value, which would suffice if we could find a pub with good accommodation. So yes, there should be ample funds. We outline the proposition to Phil, wait for him to get over being gobsmacked, and ask him to start looking for likely pubs.

In the meantime, I have a job on my hands obtaining early retirement. The sulphur project is the problem, the Directorate are expecting me to get this off the ground and are reluctant to let me go. A bit of trading is necessary. I offer to act as an in-house consultant to whoever does lead the programme until we find a pub and then to continue the consulting at a distance. This proves acceptable and my retirement is put in motion, with a contract for a three-year, off-site consultancy.

As Phil's search progresses in the Berkshire/Oxfordshire region it becomes clear that the pub he is hoping for is going to be above our heads. We need to be looking in less-urbanised towns.

I have an idea which I put to Phil and Paul. It is met with some coolness but as I push they agree it could be a solution. Patricia stayed in Exmouth after graduating and met and married David Carne, a Devonian teacher graduate, in

1975. They marry in West Hanney but live in a flat in Exmouth and then move to a house in Exeter. Over the following five or six years I help them on and off with the much-needed modernisation of the house. In the process, Paul and I get to know Exeter quite well. I propose a visit to Exeter and a scout round the area to look at any likely pubs.

There are very few on the market, but two in particular are of interest to Phil and within our financial limits. One is in Topsham, a village a few miles downstream on the Exe, and one in Exeter. Although the Topsham pub has very nice accommodation, Phil decides there isn't much scope for developing the trade.

Barts Tavern, however, is sited within the city limits and the trading books support Phil's expectations. It has the capacity for functions and the accommodation will be adequate for starters. It seems to be just what he's looking for. After the three of us bandy this around for a few hours we decide to make an offer. It requires a bit of haggling, but we hold fast and the proprietor climbs down. The pub is tied to Whitbread Brewery for supplies, it seems many pubs are tied to a brewing company in this way, and there is a ninety-nine-year lease. We miss out on the Christmas and New Year's Eve trade and move in on 4 January 1986.

Lindsay and Robert stay put in Oxfordshire, having gone into house-sharing in Grove, a few miles from West Hanney. First one and then the other has partner problems and both end up in Exeter. Richard's first marriage doesn't work out and he joins us in Exeter. When Steve leaves Harwell to work for Deltest he and Liz move to Verwood. Then he and Rich link up and he travels between Verwood and Exeter. This means that during the week all our children are round us. Even my Mum moves to a McCarthy and Stone flat by the Exe bridges after Dad dies.

Some things don't stay static for long. In 1991 Rich marries Christine and moves to Abingdon, where Chris lives with her two children, Sian and Rachel. Steve and Rich have gone into partnership to create a software business, Modular Concepts, which they develop into a company called Grapevine.

The gradual migration of my branch of the Sambell family tree from London's East End to Exeter in the South West, via Reading and West Hanney, brings us full circle, following Grandpa and Johannah Sambell's migration from the South West to the East End.

Fragment

'All are but parts of one stupendous whole' Alexander Pope, 1733

In these parts the soil is fertile red

Stubborn clay, red as rust

The fields in these parts are rarely flat

Craggy hills, scoured valleys

In these parts the streams run rapid

Cold as steel, dancing, chuckling

In these parts the Norsemen pillaged

Fought and won and settled

In these parts my forebears dwelt

Farmers, mariners, parish priest

They live no longer in these parts

As fortunes waned they drifted eastward

In these parts the past is long gone

Hope is strong, ambition thrives

The tide is turning in these parts

New towns are emerging

Families are settling in these parts

As fortunes wax the drift is westward

Exeter, the county town of Devon, was occupied by the Romans in about 50BC; they called it Isca Dumnoniorum or simply Isca. At the turn of the first millennium the Danish king of England, Sweyn Forkbeard sacked Exeter. The settlement later developed into a thriving port.

Curiously enough, the name Dumnonii, the tribal nation that lived in Devon and Cornwall, brings me right up to date. Because of my interest in rugby union I have supported Exeter Chiefs, with their logo of a Native American in flowing headdress, ever since my arrival here. The growing social pressure on racial issues has resulted in the club's board carefully and exhaustively considering the need to change the image. They will, of course, still be called the Exeter Chiefs, but their logo now will be the head of a Dumnonian warrior clad in an Iron Age helmet.

If you were to walk down Fore Street towards the Exe bridges and turn into Bartholomew Street West you would come to Barts Tavern on the right-hand side. The façade is Grade II-protected, as is the rear, which is part of the old Roman wall around the city. It is a double-fronted pub, with a skittle alley beyond the bar and the beer cellar beyond that. There is a bottle store and a walk-in fridge alongside the cellar. On the first floor there is a function room with a bar and lounge. A large kitchen and an office are also on this floor. On the second floor is a sitting room, two bedrooms and a bathroom.

In the scheme of things, Phil and Paul are joint licensees. Phil, with his background in the entertainment industry, oversees the whole shebang. Paul takes on the job of head cook and hostess. And moi? After a twenty-year interval I find myself taking early morning deliveries of beer and doing morning shifts behind the bar. Later on, Trish becomes our part-time bookkeeper, now with two children, Richard and Tamsin.

Bang bang bang.

My arm shoots out and I grab the alarm clock thinking I've overslept and the dray is here. Four-twenty.

'What the Devil,' I mutter.

Paul gives me a nudge in the back and I force myself up. Staggering downstairs I switch off the alarm and hold on to Lara as I unlock the side door.

Bang bang bang.

'Alrighty, I'm here,' I yell, and there before me is a fireman.

'Sorry about this. But you need to get prepared to evacuate. Ottens warehouse just round the corner is on fire.'

'What! But we've hardly been here five minutes.'

Four days actually, and a puerile response, but surely a natural one. The fireman tells me that from the turntable ladder they can see we have a flat roof and the wind is blowing smoke and embers in our direction. He wants to go up and see how vulnerable we are. I usher him in and lead the way to the roof garden. I can smell the acrid smoke now and the fireman tells me they will definitely have to keep an eye on the tarmac roofing from the ladder over there. Even as I follow his pointing finger, a flurry of sparks belch from a window and swirl towards us. I leave him to it, telling him I need to get the others moving. In no time Paul, Phil, Lara, the two cats and I are bundled up and ready to leave. About two hours later the wind shifts round and we are given the all-clear. Paul breathes a sigh of relief and suggests to Phil we should get tea and sandwiches on the go for the firefighters. I have a flashback to the Dunkirk Miracle, code name Operation Dynamo, and Dad telling how he ferried Mum to Dover, the car loaded to the gunnels with food and drink for the rescue crews and returning soldiers.

It's strange how memories are triggered like this. It is also odd how an old saying can be confounded, 'it's an ill-wind that blows nobody any good' is one. We'd hardly opened for business but found ourselves making headlines in the local press. Not at all how we'd planned to advertise our presence, but it served the purpose.

We inherit a barman who appears to have been running the pub. He informs us that his previous boss has a licensed hotel in Dawlish, the drinks for which were supplied by Whitbread but delivered and billed to the pub. The books would show them as consumed at Barts. One of the barman's jobs had been to transfer whatever was needed to the hotel. Apart from seeming to bolster the cashflow at Barts, the trick is that an hotelier can charge a lot more, especially for wines and spirits.

'Ouch! Have we been shafted?' I ask Phil.

'It would seem so,' he agrees. 'I think we're about to find our volume of trade is going to be less than we thought.'

We quiz the barman and discover it will have little effect when averaged out over a year, and decide to just get on with it. It does mean that we have to give the barman notice before long and run the bar on our own.

Another drain on our profit margin are the skittlers. They use the alley twice a week and expect to be supplied with sandwiches on the house. This isn't such a loss if they drink, but they don't. Phil's acumen and Paul's cooking begin to make a difference though and trade starts to improve, to the extent that, although the skittlers leave when asked to pay for their food it allows the alley to be furnished with tables and chairs as an additional bar area. Another thing that has to go by the board is giving tick to those regular customers who are apparently hard up. This could be lived with if they were to clear it down at some point, but they don't; they move on elsewhere when confronted with an ultimatum. This is a big relief to me since I have yet to be hard-nosed when such situations arise.

Our next confrontation is with the bikers. They arrive one Friday evening when a gig is in full flow. The front door access to the function room stairs is wide and hooked back. Three bikers simply ride into the hall, making an ear-splitting racket as they rev their engines. We do have a doorman, but this is well beyond his remit. It appears they have intentions to make us their rendezvous and it takes Phil and half a dozen regulars to get them out and threaten them with retribution if they show up again.

During the back end of the football season in 1986, Exeter City Football Club has a home game with Plymouth Argyle. This match is a local derby and on their way through town supporters of both clubs stop at Barts for a top-up. The usual banter gets a bit ugly when regulars get elbowed out of the way. In the resulting fracas some of the bar furniture is damaged. Lesson learnt, the pub will close during this event in future – it's only once a year. We have an unemployed furniture restorer among our clientele and we give him the job of putting the stools and tables to rights. These adjustments to our way of conducting the business just about stop the leaks in our turnover and we start to feel a bit more confident about making a go of it.

There is an off-room area in the bar in which a pool table is sited and is a focal point for some of the more active customers. When Richard joins us he naturally takes an interest in the game and we now have a pool team in the Whitbread

Devon Pub League. Needless to say, it's not too long before he has put together a team capable of winning the league.

One of the team is a tall brash man in his twenties, with a Native American hairstyle and lean regular features, apart from an oddly skewed nose. It takes some persuading to alter Spike's preference for the haircut and his attitude towards those he beats, but the nickname sticks.

Nicknames are the norm and I soon acquire the cognomen Radwaste Ron, or simply Radders, when the customers learn of my connections with atomic energy. Paul is a natural in this environment, embracing the patrons along with our children, and is Mother to all who identify with Barts. Philip? He is Boss, the ambitious thirty-two-year-old with all the answers when it comes to disputes, arbitration or the need for swift action, all achieved in a low-key, conciliatory manner.

As trade improves we employ bar staff during the day as well as in the evening. Paul's contribution in the kitchen also requires extra staff since, in addition to bar meals, there are functions such as wedding receptions, twenty-firsts, wakes, regular lunch meetings of the local Chamber of Commerce and Round Table charity affairs. The Round Table has regular monthly lunches which as many as forty members might attend. This last event has to be carefully orchestrated since there are members who have limited lunch breaks.

The function room suite is now called The Loft, a gay venue, and is open during the evenings to serve drinks in a more comfortable environment, which is shattered on Friday and Saturday evenings by gigs organised by one of Phil's friends. Ben is a lively young man with a finger on the local preferences for musical entertainment. He arranges bookings with some notable celebrities. Unfortunately, we sometimes run foul of the Noise Abatement Regulations and I find myself at odds with Ben when he disregards this law.

Two events occur which are of considerable benefit to us, and presumably to the pub trade in general. In summer 1989 an Act limits to 2,000 the number of tied pubs that brewers and chains can operate. Whitbread lets our tie lapse when the deadline passes and we became a free house; 'free', not free, meaning we can buy drink elsewhere if we wish. Second, opening hours are rationalised, allowing pubs to stay open from 11 am to 11 pm.

It appears to be a given that in a pub at the centre of a community you have 'characters', and Barts was no exception. Although we are in a city, we are just off the throughway to the centre so our customers tend to be a disparate collection of regulars from all over. Our resident furniture restorer, on benefits, is such a one. He lives about two miles away and gets around with a walking aid. This is Speedy Pete, so-called since, despite his obvious limp, he moves with alacrity when crossing busy Fore Street or whenever someone is 'in the chair'. There isn't much of him, thin with sparse hair, even his pullovers are nearly see-through. He is the manager and drummer of a local group that occasionally performs in The Loft. And, of course there is Spike

One of our customers has a small printing business in Stepcote Hill, on the other side of Fore Street. Against the conditions of his rental Rick has installed a bed under his workbench; during a chat with Phil he offers to do promotional printing for us in exchange for the use of a bathroom. This unconventional business arrangement suits both sides. It's becoming a habit for this thirty-something, coppery-haired giant to have a liquid lunch of two pints of Guinness. He's quite affable but a bit short of patience, a boxer in his spare time.

As I replenish the chilled bottle cabinets, a non-regular appears at the bar and orders a drink. Sally, a young good looker, a favourite with the customers and street-wise, is behind the bar and as she is pulling him a pint he perches on a stool in the angle where the bar top meets a wall. There are some audible intakes of breath and exclamations from others at the bar, causing her to swing round.

'Excuse me,' she says, 'but you can't sit there.'

'And why the hell not?' the punter queries sharply.

'Cos it's for Rick's bum only.'

'I don't see a label saying that.'

'Bash him,' Spike mutters.

Whereupon Basher, as she will now be known, leans across the bar, gives the offender a slap on the head and tells him a label isn't necessary as he is about to find out. Our printer has just walked in. He approaches the punter and stares him down. Basher explains that the idiot wouldn't listen, but the transgressor is up and gone. One of the regulars scribbles on a bar mat 'for Rick's bum only' and places it reverently on the stool. So, Rick's-Bum it is. Two nicknames for two delightful characters right out of the blue.

Our customer spectrum changes. My morning stints behind the bar carry a message to the effect that 'I respect your status until you come it'. Over the last few months those who do 'come it' are refused service and the ambience during the morning is now one of bonhomie. A regular and a favourite of mine is Alan. He and his wife have a vegetarian and vegan restaurant in New Bridge Street named Brambles. He is slight of build, bespectacled, with thin, slightly worried features. Frequently he would declare war on the world at large about something he'd read or seen. My many years of verbal ding-dong with Paul makes me a good Devil's advocate and between pulling pints I deliberately take an opposite view – it entrances the listeners. Brambles knows the score now and we make a creditable double act. This is naughty of me I admit, but there is a chemistry at work here; these debates draw others in who might never have had a word to say otherwise.

We have a smartly dressed regular with well-barbered, fair, almost white, hair who is a bit of a conundrum. Two Way is a lunchtime customer who likes to chat but it's hopeless to attempt a conversation because he is in and out of the door like a cuckoo. He carries a brick-size walkie talkie with him and although it has a telescopic, yard-long antenna there is no signal indoors so whenever the green light blinks up he jumps, grabs the two-way transmitter and bolts for the door. This signal deficiency prevents us from ever knowing what he's up to.

Two Way has just bought a drink for the woman sitting at a window table; a tall brunette with an eye for striking dress colours who is celebrating her fortieth birthday. She is one of the regulars Paul calls on when extra pairs of hands are needed in The Loft. Whenever she has something to say from across the room it is always with a strident ,'By the way', hence the nickname By The Way. This is okay with her … providing you remember to add on Angie.

At the back end of 1986, Paul, Tricia, David and I decide to look for a house we can share. We all need more space and buy a large three-storey semi in Sylvan Road, a mile or so from the pub. A number of modifications are needed and Richard earns his keep converting a first floor room into a kitchen/diner and the attics into two bedrooms, the latter for Tricia and David's children. Three events mar our tenure:

1. David leaves abruptly for pastures new
2. Paul's engagement and wedding rings are stolen by a walk-in burglar along with her car keys – the young blighter gives me a salute as he drives away. The car is later found unscathed in a car park but the rings are gone forever. Fortunately, we have photos of the rings and they are replaced with the insurance payout. Although these are very similar we both deeply regret this incident
3. The loss of Lara, who becomes diabetic and is going blind. She has a severe fit and the vet attending says she won't recover and should be euthanised. We reluctantly accept this and I get down and put her head in my lap. I shall never forget the feeling of something leaving us as she passes away.

Salute to a Samoyed

Our trust in you is total
We court your companionship
When the fireside is favoured
I like the way you look at me
Tail waving, white teeth tugging
At the leash if I linger on the path
And proud of the way you hold your head
Cocked as a cat arches by
Our love for you is deep and enduring
I'm holding close the ache of your passing

Lindsay has found herself a boyfriend, Julian Cole, Jules to his friends. He is a graphic designer by profession and produces brochures, programmes, magazines and so on. He is genuine Devonshire, born into a family who own a general store in Okehampton, on the edge of Dartmoor. He shares a suite of offices with

Stephen and Richard in the Real McCoy Arcade, just round the corner in Fore Street. Linds' and Jules' friendship blossoms into courtship and they marry on 6 May 1989. We hire the function facilities at the Mill On The Exe. A spit-roast pig is a feature of the reception and we receive lovely letters of appreciation from Julian's family. The newlyweds take on a studio near the Odeon, at the far end of Sidwell Street, where Linds learns the business. They own a house on Cowick Hill and it is there they add Emma and William to my flock.

Time flies and in 1996 we celebrate our tenth anniversary as pub owners and decide it is time to draw stumps. Phil is looking at other ways of using his skills, Paul and I are looking forward to a bit more travel. Whitbreads take Barts back and promptly sell it to a property developer who converts it into student accommodation. However, they can't alter the façade and the rear Roman wall is off limits. This frustrates their plans considerably and I feel Barts has a last chuckle at the outcome.

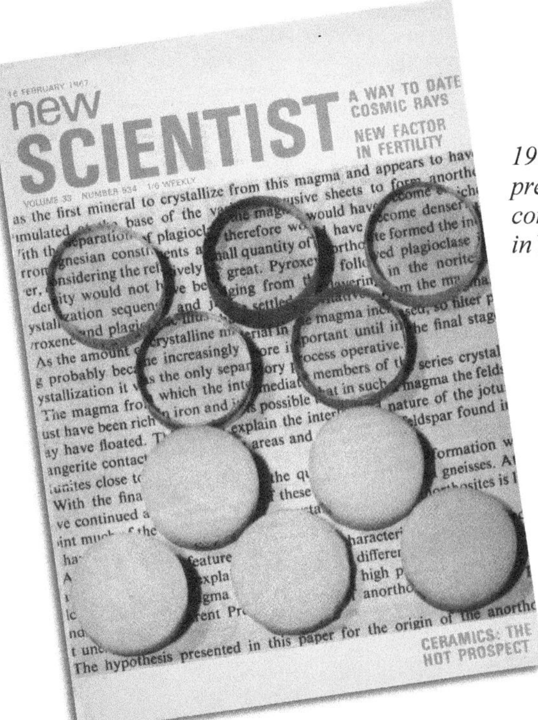

1967 'The effect of vacuum hot-pressing on the transparency of compacted magnesium oxide powder' in New Scientist *on 16 February*

1982 Me explaining to Sir Peter Hirsch (UKAEA Chairman) Harwell's research on the use of modified cement-based matrices for the immobilisation of medium-level beta radwastes. This is an EU-funded programme

1974 No camera, no lens, just a pin hole in a cardboard box, a piece of long exposure film and fifteen minutes of bright sunshine at the Song

THE SCOUT ASSOCIATION 25 Buckingham Palace Road, London SW1W 0PY

We are grateful for your help, and we want you to accept our 'Thanks Badge'. It is only a token, but it conveys the gratitude of many Scouts.

William Gladstone
Chief Scout.

1978 Presented to me on my retirement from Hanney Group by the Oxfordshire County Commissioner

1980 Champion vegetable gardener, over 50 entries. Plaque presented by the Director's wife

1980 At Song of Summer. There were four in a bed and a middle one said 'hard luck Robert'

1982 The spoof of Saturday Night at the Palladium

1981 The curls are still there...if I let them

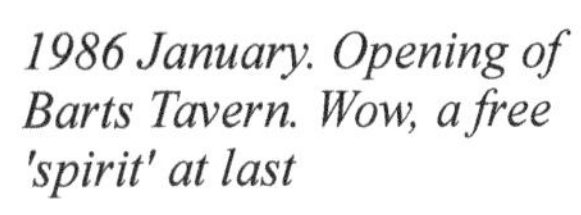

1986 January. Opening of Barts Tavern. Wow, a free 'spirit' at last

EXETER Leader

Buckerell Lodge
Where you can enjoy the luxurious surroundings of one of Exeter's leading hotels
BAR SNACKS FROM 80p
Come and visit...
THE LODGE BAR
BUCKERELL LODGE, CREST HOTEL
TOPSHAM ROAD, EXETER 52451

Exeter's Colourful Free Newspaper 45,000* copies carefully delivered weekly.

Thursday, January 16 1986

Baptism of fire at Barts Tavern

FOUR days after taking over Barts Tavern in Bartholomew Street, licencees Phil and Paula Sambell thought their pint pulling days were already over.

Mr Sambell and his mother moved to Exeter from Oxford, where he had been involved in restaurant and hotel design for one of the world's biggest hotel groups.

They bought Barts Tavern from Whitbread on a 99 year lease and were looking forward to a good trade.

Last Thursday night, however, Mr Sambell thought he heard a noise coming from the bar. He went to investigate but found nothing.

Within an hour Fore Street had been completely sealed off and over 150 people evacuated.

Exeter's largest fire in eight years had broken out in Otons warehouses, only a stone's throw from his pub.

As firemen battled to put out the flaming inferno, a weary policeman knocked on Mr. Sambell's door and told him not to be alarmed as the fire was nearly under control.

Luckily the 13 fire appliances and over 70 men managed to keep the blaze under control.

After the drama, Mr Sambell opened his skittle alley and treated the firemen who had saved his pub to hot drinks and bar snacks.

He said afterwards: "It was the least I could do."

LOANS FOR ANY PURPOSE
●Home improvements ●Furniture ●Holidays ●Car Purchase ●Free Life insurance ●Sickness, Accident & Unemployment Protection (optional) ●Loans secured on property ●No Fuss ●Settle loan anytime ●Confidential Broker Service ●Tax Relief where appropriate ●Local representative will call on request ●Self-Employed up to £7,500 without accounts.
APR 18·5%
NO FEES

1986 The 'free advert'. It differs somewhat from what really happened

1999 Brookside, You'd think the job was done

1997 Me as Santa at Exeter Rover

1991 Family group at Sylvan Road. Everyone is there, including my Mum (front left). Partially hidden is Steve's wife Liz. Trish is holding Emma

2001–04 Bowling trophies

2007 New Year's Eve Tenerife. With minder Trish. There's still life left in the oldies yet

2003 Brookside, job done. Well, it is now!

2009 The celebration of Angela's 'Silver Fox Award'. In the front row from second left are Trevor Herman (Angela's 1970s ASL), Angela, me, Trish and Angela's husband Lionel

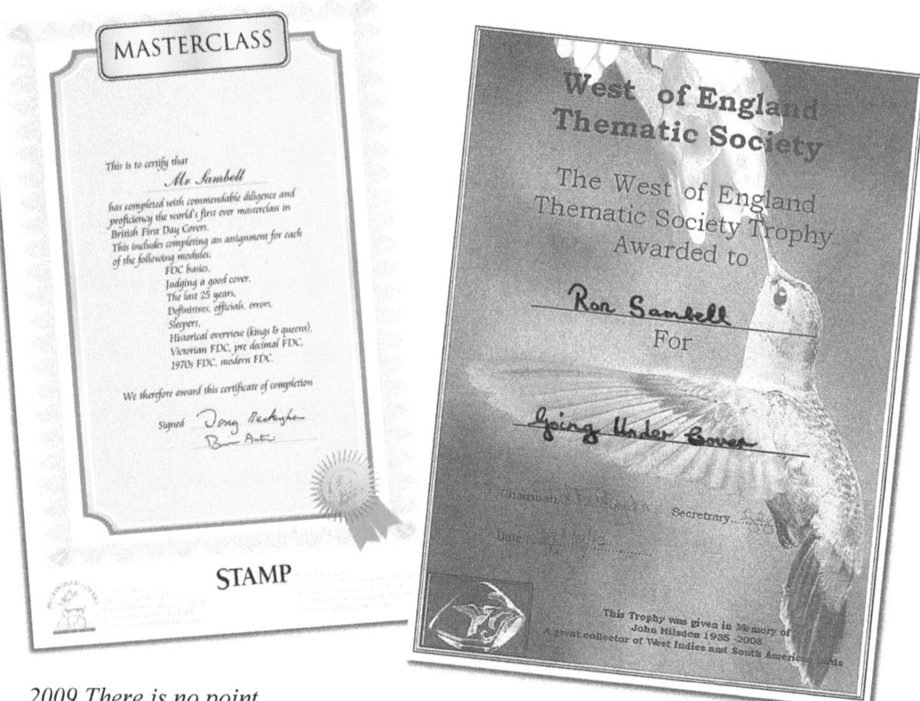

MASTERCLASS

This is to certify that

Mr Sambell

has completed with commendable diligence and proficiency the world's first ever masterclass in British First Day Covers.
This includes completing an assignment for each of the following modules:
FDC basics,
Judging a good cover,
The last 25 years,
Definitives, officials, errors,
Sleepers,
Historical overview (kings & queens),
Victorian FDC, pre decimal FDC,
1970s FDC, modern FDC.

We therefore award this certificate of completion

Signed

STAMP

West of England Thematic Society

The West of England Thematic Society Trophy
Awarded to

Ron Sambell

For

Going Under Cover

Chairman .... Secretary ....
Date ....

This Trophy was given in Memory of
John Hilsdon 1935 -2008
A great collector of West Indies and South American birds

2009 There is no point being ill-prepared if I am to take First Day Cover collecting seriously

2012 My award for winning this competition, twice. There were 38 entries for this one. My entry is still available to peruse

Chapter 11

New Start, New Boots

Looking back over my story I see now that my life happened in well-defined chunks, over which I had little control. However, I did take two crucial actions: falling in love with Paul led inevitably to the first one, when I asked her to marry me, an ex-student, ex-squaddie with virtually no prospects; the second was pure blind luck when I said 'Yes' to working at Harwell, after Freddie gave so little info and the wage was less than I needed.

The decision to sell the house in Sylvan Road is made two years after leaving Barts. It is too big and Trish and Tamsin want their own space. A good solution emerges when we find two local properties in the same road in Beacon Heath. Number 3 Brookside Crescent is a detached, two-bedroom bungalow and just right for them. Further round the Crescent, 54 is a semi-detached bungalow. Paul and I decide to make a bid for it. It has been empty for a year or so and the owners live abroad. They are concerned to get it sold. Built in 1953, it needs a lot of modernising and the long garden, a fifty-foot drop down to the brook, is neglected; but it all serves to get the asking price down by 10%.

We make our moves in spring 1998. One problem is the size and amount of furniture we'd acquired and, much as we cherish it, some has to go and we have to buy pieces more suitable for the bungalow. There is an add-on conservatory at the rear and our first decision is to make it part of the living space by removing the rear wall of the house to create an open plan lounge/dining room/kitchen area. The drive is shared but our garage is in bad shape and we decide to demolish it and have hard standing at the front; in any case, the width of the garage is such that having driven in our Honda Aerodeck you can barely open the car door. Because of the sharp downward slope the back door is accessed by a flight of concrete steps and, in due course, I remove this as I set about landscaping the garden so that the rear approach is from a series of terraced slabs. Half-way down the garden an enormous conifer overshadows everything and we enlist the aid of Stephen and Richard to fell it. Over the course of the first two years I create three further terraces, negotiated via a series of cross-flights of rural steps with wooden posts and hand rails.

In between the landscaping, we make up a bit for lost time with the family. Travelling figures high on the itinerary and we have group holidays in Europe. One of the more adventurous was spent in the Auvergne region of France, major city Clermont-Ferrand. We rent a two-cottage holiday let from a local farmer in the more southerly department of Haute Loire, together with Trish and her two children, Richard and Tamsin; and Richard, Christine, Jack, Stephen and Elizabeth. The River Allier, which has deep gorges and is notable for kayaking, is close by. Stephen demonstrates his great ability at this sport. Imagine my chagrin when I end up going backwards down the sluice at the start, not helped by the others cheering. Tricia's problems begin when we enter a stretch of the river that is shallow and rocky and relatively fast flowing. Richard has to jump out of his kayak and steer the nose of hers out of trouble. Did we have safety gear on? Probably, but I can't remember. On this holiday we see Jack at the wheel of the farmer's tractor – was this the beginnings of his career in industrial vehicular design? This is a mountainous region of France, the Puy de Dôme being the highest of these extinct volcanoes at 4,806 feet. We climb to the top after parking the cars part way up and visit the ruins of a Gallo-Roman temple of Mercury, built in the second century and excavated in the 1880s. I have to confess, I was much more interested in getting my breath back.

Our love of Monet is fulfilled by visiting his gardens in Giverny with Trish. We were cock-a-hoop standing on the bridge and looking across at the water lilies. We made a point of visiting Avignon and its half-bridge, well known thanks to the nursery song 'Sur le Pont d'Avignon'. It was built in the twelfth century across the Rhône, but because it was damaged so frequently the town folk gave up in the sixteenth century. We walked out on the remaining part.

Paul and I also travel alone and we stop for lunch at a lake just north of Rome. I find a nice example of pumice stone when I go for a paddle – I still have it. We make a point of revisiting Rome to explore the Colosseum, the world's largest ancient amphitheatre. The wall you first see as you approach from the Forum is actually the inner wall, only a small section of the outer wall remains, the rest having been filched by local builders. Looking down on the arena I experience a frisson from 'someone walking over my grave' as I picture the carnage which must have occurred.

While we are in southern Italy we visit Pompeii. In 79 BC Vesuvius erupted and this prosperous city was suddenly smothered in volcanic dust and pumice to such a depth that it disappeared from the map. Excavations since the 1800s have revealed many examples of people perfectly preserved in the attitudes they were in at the time. It feels a bit odd to be walking in ruts in the roads from so long ago.

Trish is with us in Malta for the run up to Christmas. What stays in my mind is the abundance of grottos filled with detailed miniaturisations of the life of Jesus. In contrast, our holiday in Cyprus is filled with visits to museums and architectural relics adorned with murals of Greek mythology.

Most of this travelling and holidaying is in the years after leaving Barts and includes visits to Lindsay and Julian in France. They now reside permanently in Le Breuil d'Haleine, a small village about eighty miles east of La Rochelle, just a few minutes walk to the Charente river. From here we visit the historic towns of Poitiers and Angoulême. The nearest shops are in Civray, a few miles away.

By 2002, the years of stop-start slog in the garden are behind me and we decide to have a pergola on the patio further back from the house. There must be a bit of my Dad in me because I enjoy working with wood. Stephen has been in touch to say he will be kayaking on the Teign on Sunday and would like to stay with us on the Friday and Saturday. We are pleased and I suggest he comes prepared to help me erect the posts and rails along the flights of cross-terrain steps; you have to take these opportunities when they arise. The weather is perfect and the posts are concreted in place in no time. We sit in the garden, having a long chat about his separation from Elizabeth and the protracted divorce. Puffing away on his pipe he seems resigned to the event and tells us he has a new partner, a lady he knew from way back and who he bumped into in Abingdon, where he and Richard now have their offices. Bright and early on Sunday we wave good-bye to him. In bed that night Paul and I mull over our children's failed unions.

I wake to another beautiful morning. The phone rings. It is Richard. He delivers the hardest message either of us have ever had to deal with. Stephen has died. Paul and I draw together and hold on tight, as this terrible news sinks in. With a long, long sigh that becomes a wail, Paul sags in my arms.

After completing a kayaking stint which finished at Bantham Beach near Kingsbridge he had a massive heart attack. Fellow kayak medics did what they could but he had already gone. This date, 12–10–2002, will never be forgotten.

As I sit in the chapel listening to the tributes to Steve from family, friends and associates I find some comfort in the extent to which he loved and is loved by these people and is missed by them. There is even a homily from a schoolmate who had only just recently contacted Steve to renew their friendship.

Stephen had done well at Alfred Sutton Primary, including excelling at games and athletics. He went on to Stoneham Boys Grammar, where he joined the Army Cadet Unit; getting his finger jammed in the breech of his rifle cost him a GCE maths result. He too went through the Scouting ranks, eventually becoming a Queen's Scout and Troop Leader. Although he did well in both class and outdoor activities he couldn't leave school fast enough. In fact, he told me he hated it. Music became an abiding love. He learned to play the 'cello and performed with the Reading Youth Orchestra, before turning to the classical guitar. Much later, when he lived in Verwood, he formed a kayak and canoe club and reached national status in competitions. He excelled in chemistry as a discipline, and after a spell in a paint development laboratory in Slough he followed in my and Richard's footsteps to work at Harwell in the Chemistry Division, where he discovered his gift for designing computer software programmes.

As the millennium looms I feel restless and at a loose end intellectually. Far back in my childhood I'd started collecting stamps. Because of his developing role at B&H Dad regularly received orders and invoices from other countries and he would bring the envelopes home and Zeb and I would soak off the stamps. He bought us albums in which to put them but the war interrupted this hobby, although we both kept the albums. Among our customers at Barts Tavern was a stamp dealer with a shop in Princesshay. In due course, he asked if I was a stamp collector, to which I gave the well worn reply, 'Yes, when I was a boy.' He nevertheless proceeded to show me two things, a 'penny black' and a set of four, what he called, 'first day covers'. The penny black was pricey but he showed me a catalogue produced by Stanley Gibbons with their value for a penny black printed from the same die as the one he had. With his dealer's hat on he pointed out the good features of the stamp. His asking price was 50% less than theirs so I decided to buy it. The covers were a different kettle of fish altogether. They had been produced by Benham, a dealer in Folkestone, in 1980. Each cover (envelope) had one of a set of four bird stamps on it and each stamp was cancelled with a postmark showing the first date of issue of the stamps. Apart from the high quality of the stamps themselves, which were commissioned and produced by Royal Mail, two things stood out: each cover had a different illustration relating to birds and each had a different postmark, which was also connected in some way to birds. This set was the first of a series produced by Benham, who called them 'small silks', by virtue of the feel of the illustrations and the relatively small size of the envelopes. Benham continued to produce these covers for every special/commemorative set of stamps issued between 1980 and 1996. The dealer was asking £1.50 for the set he showed me. I bought it and placed an order with Benham for back and future issues. I have all of them. It is one of the reasons why I decided to take philately seriously as a hobby. I joined the Exeter Philatelic Society in 1999, which met monthly. The main advantages of taking this step are accessibility to advice and the opportunity to buy from club members or visiting dealers.

There are four main sections of philately: Stamp Collecting per se, Postal History, Thematics and First Day Covers (FDCs). I have a strong interest in all four and have used Thematics/Postal History displays to produce prize winning entries in West of England Thematics Society competitions.

There were three members of the stamp club who lived in my area and we car-shared. This was particularly useful when visiting other clubs in Devon. Andrew, a car-sharer, who lived alone just up the road from us, contracted motor neurone disease, complicated by a heart problem. His health deteriorated steadily until he was in need of palliative care. It appears he was averse to hospitalisation and his care was put in the hands of Exeter Hospice. Paul or I visited regularly and we became very impressed by the quality of care of the hospice. I was already aware of their activities because they have an on-line shop which often has philatelic material for sale. Andrew's end of life care was inspiring and from that time on we have supported the charity by subscription. All income from sales of this book will be donated to Exeter Hospice.

Browsing at a Philatelic Fair in London I come across a dealer who specialises in autographed covers. One carried the autograph of Bobby Moore, the 1966 World Cup captain, on a FDC commemorating the England victory. Since I still have my memorabilia of the match I decide to buy it and go on a philatelic hunt. I now have the other ten autographs of the team, all on one FDC. The autograph of Alf Ramsey, the team manager, is on a separate cover. I catch the collecting bug for sporting autographs on covers and most sports are represented in my collection.

This autograph lark takes a serious hold on me and, in order to finance it, I look around for part-time jobs. One of our erstwhile Barts' punters is the manager of the Rover Group car franchise, situated on the edge of Sowton Industrial Estate on the Honiton road. 'The Only Major Car Dealer NOT On Marsh Barton' the billboard says. On the off-chance, I call in and enquire if he has any jobs going. It is August and he is short-staffed due to holidays.

'Yes, can you start tomorrow? In fact I can take you on full time.' He has just sold a new MGF to someone moving up to Glasgow and the owner requested someone drive it up there to get a good chunk of the running-in done and to drive back an original 1995 MGF in a part-exchange deal. I promise to let him know by tomorrow.

'What do you think, Paul?' I ask when I get home.

'If it takes the edge off your collecting costs why not. But for starters you can take me with you to Glasgow.'

So, on a bright summer's day in 2001 we set out on the 450-mile trip to bonny Scotland. For the next four years I travel all over the South West delivering and collecting cars and parts.

Christmas 2003 and the manager calls me into his office.

'As you know, we are putting in a special effort to attract buyers next Saturday. Now, it's not in your job description, Ron, but I'd like you to be a Father Christmas for the day, okay?'

I demur, to begin with I don't work Saturdays. Second, I'm not the right build. But, most importantly, I just don't have the jovial gusto for the part.

My boss negates all this, 'Come on, Ron. I remember how your bonhomie was so important at Barts. It might have been out of character for you but it came across as natural. You'll have no difficulty switching on for the children.'

What could I say? The guy was almost pleading.

Two more activities begin to occupy my time: bowls and bridge. I start from scratch with both. They come about when I take a part-time job as evening dogsbody at the Exeter Arena, also to fund my collecting. Both games are catered for at the Arena and I just have to give them a go. I represent the bowls club in the Devon League and win the club singles, mens pairs and triples competitions.

Then, on Christmas Eve 2004 I'm carted off to Royal Brompton Hospital in London for open-heart surgery. Although the procedure goes well I have an allergic reaction to the morphine administered as a pain killer and endure six days in intensive care, where hallucinations and delusions come thick and fast. These seem very real indeed and are still lodged in my mind: I'm in a ward in a church tower and am rescued by a fireman when the building catches fire; along with the nursing staff I've been selected to be in a film titled *Leaf* and get really agitated when the staff cause interruptions due to shift work; I'm in a ward on the Thames embankment with big double doors and radioactive waste is dumped straight into the river from lorry after lorry, which results in rapid growth of algae that dams the flow and I lie watching the water rise to floor level, I try to tell a doctor of the danger but he just shrugs and puts up a placard advertising drinks and snacks; there's a room full of televisions and I am convinced this will allow me to talk to my family; I daren't look down at the floor because I see hordes of beetles scampering towards the door.

Once the allergy is recognised the prescription changes but it still takes a week for the drug to leave my system and for me to start taking notice. Richard and Chris pay me a visit (no illusion) on New Year's Eve while they're in London to see the jollifications. They have told me since they thought I'd gone bonkers. I come home ten days later and three stone lighter, absolutely shattered and Paul has to take even the simplest of tasks out of my hands.

Rehabilitation takes about three months and although I go back to bowls my coordination is shot and I call it a day. Fortunately, my grey matter is still in place and I pick up on the bridge sessions. I am reasonably good at this card game, particularly the playing of the hands after the bidding is done. My enjoyment is due in great part to two successive partners. The first is a retired police superintendent who went to the same classes I did. A well-built, confident fellow with large capable hands and sandy hair arranged to mitigate the thinning, he's a bit of a chancer when it comes to bidding. He dies suddenly and unexpectedly. At his funeral I recognise the usher at the church entrance, Dave, one of our Barts regulars, and realise I'm being greeted by my bridge partner's son.

My second regular partner is a maths teacher who did the session of classes after mine. Hilary soaked up the ACOL bidding system easily. She does not let me slide and corrects me when I cock up the bidding – in the nicest of ways.

I'm not sure how I come to be an invigilator at St. Peter's School. Trish works there, which is probably what's behind it. Along with five or six others I walk slowly and quietly through the array of fifty desks, keeping a watchful eye on the candidates. Out of curiosity I check the number of left-handed pupils, it is seven today. My thoughts drift back to my school days and the difficulty I had using a dip pen with a nib designed for right-handers. I had to push it to write, net result – splattering. To avoid it you had to contort your wrist to draw the pen across the page. I was quite surprised to see that five of these left-handed students also contort their wrists, even though the problem isn't there when using a ballpoint.

Paul has a hip replacement in 1999 and, although it takes a while, she is able to lead a normal life until her right knee gives up. It is replaced in 2005. This operation is not very successful and as time goes by she has to rely first on walking aids and then on a wheelchair. I become aware of things not being right, not only because of Paul's lack of interest generally in me, but her disinterest in herself. She frequently forgets to take her medication, or else decides it isn't necessary to take her pills or inject insulin. A urinary infection leads to hospitalisation for several days. Early in 2007 I decide to take on her full-time care and all the household tasks. There follows a period of resistance from her and determination from me – a down time in our marriage. Eventually her increasing immobility causes her to become more resigned to our home situation. The mobility problems don't affect Paul's desire to experience foreign parts though and we carry on with our travelling and our interest in the arts.

Although I have no problem with the workload I can't resign myself to the necessary repetition of household chores on an endless weekly basis. I start investigating the pros and cons of having some home help. So, in 2008 I latch on to a local organisation and take on Sue. What a wise and fine choice. Sue is a retired nurse who can't stay idle and once a week she spends two hours vacuuming and cleaning the ground floor. She is Devon through and through and in no time at all she and Paul become gossip buddies. Her husband Graham is a bus driver, a good source of tales to tell. Sue's reporting of these provides much-needed levity. Covid-19 put an abrupt stop to the regularity of Sue's visits. The passing of Paul proves very distressing for her. But, here we are in 2022 and my stalwart housekeeper is still with me.

Since Dinglebirdy and I went our separate ways aged seventeen we've kept in touch. I learn from his family that he has died as the result of a horse-riding accident. I'm deeply saddened because, outside of family, I've had few close friends and Feliks was one of them. The following started out as a tribute to Dinglebirdy but evolved into a more personal observation on life. This exercise can generate ideas to help express a depth of feeling. It is called 'free writing'.

Free writing is where a phrase contains a word or idea which is picked up in the next phrase and so on, there doesn't have to be a storyline. It can be used by writers as a tool to kick-start a stalled mind. A focused free write is one where an overall theme does emerge. I'm not sure this one deserves a title but here goes.

A Beach In Winter - A focused free write

A beach in winter; a bleak experience coupled with thunderous noise and chaotic movement – and yet – there is a permanence and regularity in this shifting scene which induces calm, even sadness. Sadness fills mind and heart lingering here and reflecting on the loss of a very dear friend. It was an accident, a slipping of his saddle. White horses are charging up the clattering shingle; rearing, white-maned echelons of unbridled energy, the cavalry of the fabled Sea-Gods, charismatic but cruel Immortals. Is it so different in our prosaic indifferent world? A rebuttal to this ugliness is occasioned. 'Yes, it is otherwise. It isn't all like that.'

Compassion aid advice leadership compete with the more sombre aspects of humanity. Stark and dark the silhouette of a freighter hull-down on the horizon, motionless as the stars. This immobility is an illusion which the passing of time will shatter. Time was when reflection was secondary to expectations. Of these there were many: scholarship, career, status, leisure-time, family, security.

'Safe as houses?'

The tower block collapsed all the same, no one injured, no longer habitable. The rubble has succumbed to new uprisings – nature's sure response to the desolation and despair – a vigorous never ending cycle, an undulating palette of colour, a Sea of Optimism.

Time passes, with notable events to record:

- Stephen's daughter Lorna leaves university with an honours degree and marries Matt Harris; they have two children, Bryony and Oscar
- Richard and Christine have a son, Jack
- Christine's two girls from her previous marriage, Sian and Rachel, now all grown-up and married, add five lovely great-grandchildren to my extended family
- Jack and William go through university and emerge with honours degrees
- Trish sells her bungalow and moves to a semi-detached house in Sullivan Road
- Robert marries Lucy Kiss – Lucy's son Zachary is another addition to my brood
- Patricia's Tamsin gains an honours degree and a distinction PGCE in Business and IT
- Lindsay and Julian become ex-pats when they decide to live permanently in France
- Tamsin marries Zac Newton, they present me with a great-grandson Theodore, called Teddy. I draw the line, he will always be Ted to me
- Tricia's Richard marries Lucie Damborská from the Czech Republic; they live permanently near Prague; his children from his first marriage, Joshua and Hannah, remain with their mother, Becca

When Tamsin gets serious with Zac, Trish talks to us about our future needs and suggests three things: i) she sells Sullivan Road to Tams and Zac; ii) she buys out her siblings' shares in our house; iii) she funds a loft conversion in our bungalow and lives with us (the deeds to our bungalow were bequeathed to our children after the move from Barts Tavern, with the proviso that Paul and I be suitably housed until our deaths).

In 2013 all these plans come to fruition. I take a risk and have fourteen solar panels installed on the back and front of the bungalow. I also invest in a digital voltage optimiser, which kicks-in whenever items such as the electric kettle or washing machine cause a jump in load. After eight years I can report the return on these investments was an average of 12% per annum. The actual saving on electricity used in the house has been about 20% per annum. The return on the optimiser is more difficult to quantify, but I'm pretty certain it has significantly lengthened the life-times of all our electrical devices. So, all in all, a good venture.

In January 2019 I pull up outside the house after an evening at the Bridge Club and react anxiously to the sight of an ambulance. Two paramedics are with Paul and Robert, who has come round to give a hand. Trish tells me Paul has had a fall in the bathroom as she assisted her out of the wheelchair. Paul had wrenched the grip bar she relies on out of the wall and she'd gone down heavily, hitting her head and tearing the skin between her toes. By the time I arrive the paramedics have got her up and on her bed. Paul's general condition worsens as a result of this fall: her appetite falls away rapidly; her confidence when moving, even from armchair to wheelchair, is badly affected; and her usual reluctance to letting me look after her has given way to passive acceptance. The community nurses are in frequent attendance and I am given help in preparing meals for her, which are wasted. Our GP comes in regularly. Why didn't I see this coming? On the evening of 20 February he talks to Trish and me quietly about Paul's decline and suggests we gather the family because Paul is approaching her last hours. On the following afternoon our children all gather round Paul's bed and say their goodbyes in turn. I return to my seat and hold her close. She is quite relaxed as her last breath leaves her.

This phrase encapsulates the emptiness in me:

> *'My heart is not with me but with you and now, even more, if it is not with you it is nowhere'*
>
> *Extract from the* Letters of Abelard and Heloise, *translated by Betty Radice*

Paul's funeral is attended by many family and friends, not least from those far-off days at Barts. I am embarrassed at not recognising some and forgetting the names of others. My tribute to Paul is not a direct approach, because I know I will break down before uttering a word. Instead, I include in the order of service a poem which she wished read out. When Stephen died it was given to Paul by a close friend and surrogate son from Barts, who died soon after from AIDS.

Paul always considered this poem as a farewell to her, as well as a tribute to Steve.

And now here we are, with Paul, using this poem to bid goodbye.

Do not stand
By my grave and weep.
I am not there,
I do not sleep -
I am the thousand winds that blow,
I am the diamond glints in snow,
I am the sunlight on ripened grain,
I am the gentle autumn rain,~
As you awake in the morning's hush,
I am the swift, up-flinging rush
Of quiet birds in circling flight,
I am the day transcending night.
Do not stand
By my grave, and cry –
I am not there,
I did not die.

This poem was composed by Clare Harner, Topeka, Kansas, in 1834. It was published in the December 1834 issue of The Gypsy poetry magazine with the title 'Immortality' (Source: Wikipedia)

Twelve years of caring was the focus of our lives and my love. What lies ahead is something I cannot face just now. I know my family will support me as much as I require but, even though it's my ninety-first year, I am reluctant to step back. I still drive. I still walk Emma's spaniel Kailo. I still pick blackberries in Eastern Field, tend the garden, go shopping. And I still play bridge, keep the philately ticking over and, much to the relief of certain members of the family, I still make a reduced-sugar orange/lemon marmalade. So, is this my 'nowhere'? I'm certainly in limbo without Paul. And if not nowhere then somewhere, and where is my 'somewhere'?

Somewhere

You, me, everyone; all linked by many roads. Travel a route and create the Past. This path rolls up even as we tread it. There is no 'going back', the Past is Nowhere. Often the ways diverge or cross or merge. Do what we will the ways we choose determine the Future. My journey is short now; the Future so close but I can't grasp it. The way beyond will never exist. Only tread/decide. The Future is another Nowhere. Between the two is NOW – this then is my Somewhere.

Am I losing it people?

'Madness need not be all breakdown. It may also be breakthrough' R.D. Laing, The Politics of Experience

In the early 2000s I did an Open University, second-level course in creative writing (as in novel writing), but with the clear objective of writing my life story, a product of my memories not of my imagination – although I have had a stab at producing a novella but not published it. During my career I wrote many thousands of words and, indeed, some of my research was original and productive, but always based on fact and written in formal English. Perhaps the creative germ was incubated when I collaborated with Roger Davidge and when I did the sulphur thing and joined the conservation drive.

This ambition to write an autobiography languished in the laptop as I became a dedicated carer and travel companion to Paul. Now, there is a sense of urgency and a strong desire to get my writings into print. Apart from a fantastic ten-day visit to Richard and Chris at their holiday home in southern Spain in the autumn of 2019, there has been little to distract me from getting stuck in to key-tapping. Again, 'it's an ill wind … '. The coming of Covid -19 and lockdown provided me with plenty of time to get my memories in order and to verify a lot of my bits and pieces of knowledge. The many photos I have, some going back to the 1800s, are now as factually accurate as possible. The real difficulty is choosing those to include in my story.

My narrative brings me more or less up to 2022. Last year I spent a few days in Abingdon with Richard and Chris. Rich and Vicky, Zeb's second child, arrange a surprise visit to Zeb and Mavis. When he realises I'm standing on his doorstep the war-cry of 'Odin' rebounds as we hold each other, much to the puzzlement of everyone else.

I sit here with my laptop and watch a shower of rain angle flat out across what, ten minutes ago, was vacation sky. Have I captured your attention I wonder? I did warn you my story is that of an ordinary person, spread across the length and breadth of southern England as I migrate back to my paternal family's homeland. The metaphorical 'holes in my boots' are, to me at least, extraordinary things that shaped my days on earth, my character and my beliefs.

You might feel that I have not dealt adequately with your place in the story, and for this I do apologise, although you have to remember this is my life story and your turn is there to be grabbed.

The readability of this book stands or falls in a format of my choosing; I make no apologies for it. I take full responsibility for any mistakes and omissions.

Stay safe dear Family.

All my love.

Dad, Ron, Grandpa, Papa

The Challenge

My take on the two-line poem *Terezin* by Michael Longley.

TEREZÍN

No room has ever been as silent as the room
Where hundreds of violins are hung in unison.

Michael Longley Sarah Longley

Published by Andrew J Moorhouse
Holocaust Memorial Day 2020
97/100

First reactions: a sense of stillness, a sad empty room.

Then: people have departed. There are about thirty violins in a full orchestra; 'hundreds of violins' implies a departure of many orchestras – and, by extension, a massive departure of people.

Perhaps I'm being a bit unfair. WWII is well within my lifetime and I was quite aware of Terezin as a place in Nazi-occupied Czechoslovakia and its role in the Holocaust. I have no difficulty in understanding the meaning within this two-line poem. There were so many musicians in this death camp, different full orchestras could have performed each day, for many days.

Never mind if you have not been able to deal with the challenge. What I would ask you to do though is to reflect on the possible consequences of ignoring what is happening ethnically in the world today.

Acknowledgements

First and foremost, grateful thanks to my son-in-law Julian Cole for taking on the task of getting the layout of this book ready for the printers, for his artistry in designing the cover and for his skill in the quality and arrangement of the pictures.

Along the way, my five surviving children, my wider family and my friends have all contributed by confirming my memories or jogging me into yet more detail. I thank them.

I am in debt to Patricia, who rescued me time and again when I floundered (mis)managing my laptop and the printer/scanner.

This transition from recording events formally to writing something approaching a story I owe to the Open University and FutureLearn.

My family trees are more complete thanks to having MyHeritage in the background.

I owe thanks to Wikipedia and Google for verification of my bits and pieces of knowledge.

The encouragement to plod on when my confidence has faltered I attribute to Michael Oke's book *Write Your Life Story*, in particular the latter part of it which deals with the three Ps: presentation, production, publishing.

Lastly, my heartfelt thanks to empathetic copy editor Alison Shakspeare of Shakspeare Editorial.

Appendix 1 Published work

There were many Harwell reports, overviews and patents classified as secret or sensitive. Reports on commercial contracts were also restricted. I surmise that with the passing of time many will have been declassified. Here, I have only listed work published for the public domain during my time at Harwell.

1954 Alan Blainey presented a paper, 'Fluon-impregnated self-lubricating bearing materials' at a symposium on Powder Metallurgy, ISI Special Report No. 58, 1956. My contributions are acknowledged in Appendices 1 & 2, p 230: 'Variation of frictional properties of Fluon with temperature and pressure' and 'Rheological properties of Fluon at elevated temperatures'.

1959 'The uranium monocarbide and uranium mononitride systems', RAJ Sambell and J Williams, *Journal of Less Common Metals*, vol 1, pp 217–226.

1960 'The variation in unit cell-edge of uranium monocarbide in arc-melted uranium/carbon alloys', J Williams, RAJ Sambell and D Wilkinson, *Journal of Less Common Metals*, vol 2, pp 352–356.

1960 'Plasticity and fracture in single crystals of MgO', FJP Clarke and RAJ Sambell, Transactions of the VII International Ceramic Conference, London.

1960 'Microcracks and their relation to flow and fracture in single crystals of magnesium oxide', FJP Clarke and RAJ Sambell, *Philosophical Magazine*, vol 5, pp 697–707.

1961 'Some effects of thermal shock on flow and fracture in single crystals of MgO', FJP Clarke, RAJ Sambell and GD Miles, *Transactions of the British Ceramic Society*, vol 60, pp 299–32.

1961 'Basic mechanisms leading to fracture by thermal shock', FJP Clarke, GD Miles and RAJ Sambell, Proceedings of an Oxford Conference, *Science of Ceramics*, vol 1, pp 223–238.

1962 'Mechanism of microcrack growth in MgO crystals', FJP Clarke, RAJ Sambell and HG Tattersall, *Philosophical Magazine*, vol 7, pp 393–413.

1962 'Cracking at grain boundaries due to dislocation pile-up', FJP Clarke, RAJ Sambell and HG Tattersall, *Transactions of the British Ceramic Society*, vol 61, pp 61–66.

1963 'Diamonds in Atomic Energy Research – 2', RN Simmonds, RAJ Sambell and A Briggs, *Industrial Diamond Review*, June edition.

1964 'The strength of irradiated MgO', RAJ Sambell and R Bradley, *Philosophical Magazine*, vol 9, pp 161–166.

1967 'Fabrication of fully dense transparent polycrystalline magnesia', RAJ Sambell, DG Miles, J Rutherford and GW Stephenson, *Transactions of the British Ceramic Society*, vol 66, pp 319–335.

1970 'The Technology of Ceramic Fibre / Ceramic Matrix Composites', RAJ Sambell, *Composites*, vol 1, issue 5, pp 276–285.

1971 'Carbon Fibre Reinforced Ceramics', DH Bowen, DC Phillips, RAJ Sambell and A Briggs, Proceedings of an International Conference on Mechanical Behaviour of Materials, Kyoto, Japan, *Society of Materials Science Japan*, 1972.

1972 'Carbon fibre composites with ceramic and glass matrices: Part 1 Discontinuous fibres; Part 2 Continuous Fibres', RAJ Sambell, DH Bowen and DC Phillips, *Journal of Materials Science*, vol 7, pp 663–681.

1972 'The mechanical properties of carbon fibre reinforced Pyrex glass', DC Phillips, RAJ Sambell and DH Bowen, *Journal of Materials Science*, vol 7, pp 1454–1464.

1974 'Ceramics – Materials with an Engineering Future', RAJ Sambell and RW Davidge, proceedings of a conference on 'Conservation of Materials', Harwell; published in *Atom*, No 218, pp 215–229.

1978 'Technology of Pressure Assisted Sintering of Ceramics', RAJ Sambell, ch 10 in *Sintering*, edited by MB Waldron and BL Daniell, Heyden, 1978.

1981 'The effect of hot pressing additives on the leachability of hot pressed sodium hydrous titanium oxide', TM Valentine and RAJ Sambell, *Nuclear and Chemical Waste Management*, vol 2, pp 125–130.

1982 'A contribution towards a standard leach test for immobilised low and medium radioactive waste', RAJ Sambell, C Smitton and A Elsdon, *Nuclear and Chemical Waste Management*, vol 3, pp 125–129.

1983 'Conditioning of low and intermediate level radioactive waste', *IAEA Technical Report*, Series 222 (RAJ Sambell was UK representative on the international working party for this report).

1983 'Characterisation Input to the Quality Assurance of Low and Medium Level Waste Forms and Packages in the European Community', R de Batist, R Koster, P Pottier, RAJ Sambell and RA Simon, IAEA Conference on Radioactive Waste Management, Seattle, 16–20 May. English version edited and compiled by RAJ Sambell.

1983 'Characterisation of low and medium level waste forms', edited and compiled by RAJ Sambell, *Euratom report EUR8663 EN.*

1984 'Characterisation of low and medium level waste forms' edited by P Vejmelka and RAJ Sambell, *Euratom report EUR9423 EN.*

1985 'Conversion of zircalloy to a massive, chemically inert form', HA Kersey, RH Knibbs, AC Mercer, AK Nickerson, D Pearson, RAJ Sambell and RI Taylor, *Euratom Report EUR936.*

Appendix 2 Patents

Here I have listed those patents where I am a patentee. Harwell Patents Office vetted all proposed patents and had the last say about going ahead with an application.

1967 'Forming fine capillaries in solids', DH Bowen and RAJ Sambell, complete patent specification C42713, filed 21 September.

1968 'Improvements in or relating to apparatus for coating fibres', DH Bowen, NJ Mattingly and RAJ Sambell, British patent 1279252.

1972 'Ibid', DH Bowen, RAJ Sambell, KAD Lambe and NJ Mattingley. Filed in USA 1969. Ratified in 1972, US patent 3646908.

1972 'Carbon fibre reinforced grinding wheels', DH Bowen, AT Slater and RAJ Sambell, provisional patent specification P20263/72, filed 2 May.

1973 'Lead recovery from lead/slag mixtures', FS Feates, RAJ Sambell, RW Davidge and DVC Jones, complete patent specification C1918/72, filed 9 March.

1974 'Improvements in or relating to glass-ceramic fibres', J Bacon, P Knott, KR Linger and RAJ Sambell, British patent 1495022.

1975 'Discontinuous fibre spinning process', RAJ Sambell, DVC Jones, AT Slater and RF Preston, complete patent specification C32552/76, filed 6 June.

1976 'Production of fibres', RAJ Sambell and DVC Jones, patent of addition 11534 BtH.

There are no patents for me beyond 1976 after I move away from up-front innovative R&D to an oversight role across the Ceramics Centre and Materials Development Division and then move into international radioactive waste management programmes.

Genealogy and Family Trees

Introduction

I have presented direct genealogical DNA strands. They are brought together where my Mum and Dad get married. I have been fortunate in tracing nine generations in the case of one line, back to 1656 during the reign of Charles II, 'The Merry Monarch'.

The trees contain much more family history, such as more detail for births, marriages, deaths and siblings. The information contained in these trees is far from complete and work is still in progress. Verifying the data is time consuming.

DNA Lines

Two lines are shown: one paternal and one maternal. In each case there is an offspring who carries Collison or Horsewill genes. When my parents had me and Zeb, we carried both Collison and Horsewill genes.

Line 1			Line 2		
James Collison 1656 - 1737	m	Elizabeth Ringrose 1663 - 1724			
Peter Anderson 1689 - 1755	m	Elizabeth Collison 1680 - 1757			
William Ruston ?	m	Elizabeth Anderson 1721 - 1770			
Benjamin Shakespeare 1747 - 1794	m	Hannah Ruston 1748 - ?			
Benjamin Shakespeare 1775 - 1845	m	Elizabeth Matthews 1780 - ?			
Joseph Shakespeare 1803 - 1878	m	Mary Ann Jones 1805 - ?	William Horsewill ?	m	Mary Matthews ?
Jonah Shakespeare 1847- 1915	m	Hannah Hickman 1847 - 1900	James Low ? - 1882	m	Mary Horsewill 1839 - ?
Alfred Sambell 1872 - 1934	m	Johanna Shakespeare 1874 - 1930	James W. Low 1864 - 1911	m	Mary Dodsworth 1871 - 1953
Alfred Arthur Sambell 1902 - 1986			m		Hannah Horsewill Low 1904 - 1992

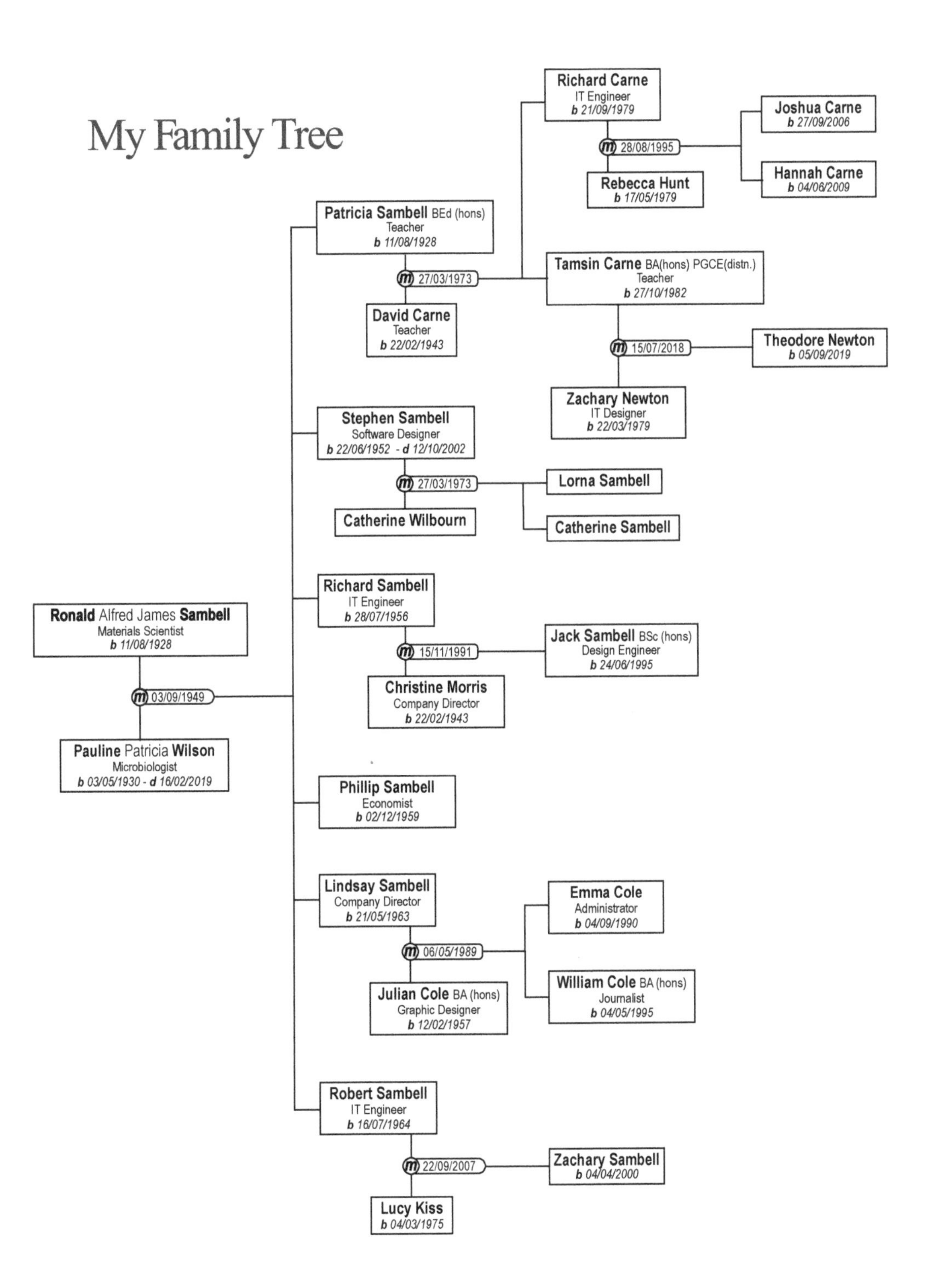
My Family Tree
Ronald Alfred James Sambell
Materials Scientist
b 11/08/1928
03/09/1949
Pauline Patricia Wilson
Microbiologist
b 03/05/1930 - d 16/02/2019
Patricia Sambell BEd (hons)
Teacher
b 11/08/1928
27/03/1973
David Carne
Teacher
b 22/02/1943
Richard Carne
IT Engineer
b 21/09/1979
28/08/1995
Rebecca Hunt
b 17/05/1979
Joshua Carne
b 27/09/2006
Hannah Carne
b 04/06/2009
Tamsin Carne BA(hons) PGCE(distn.)
Teacher
b 27/10/1982
15/07/2018
Zachary Newton
IT Designer
b 22/03/1979
Theodore Newton
b 05/09/2019
Stephen Sambell
Software Designer
b 22/06/1952 - d 12/10/2002
27/03/1973
Catherine Wilbourn
Lorna Sambell
Catherine Sambell
Richard Sambell
IT Engineer
b 28/07/1956
15/11/1991
Christine Morris
Company Director
b 22/02/1943
Jack Sambell BSc (hons)
Design Engineer
b 24/06/1995
Phillip Sambell
Economist
b 02/12/1959
Lindsay Sambell
Company Director
b 21/05/1963
06/05/1989
Julian Cole BA (hons)
Graphic Designer
b 12/02/1957
Emma Cole
Administrator
b 04/09/1990
William Cole BA (hons)
Journalist
b 04/05/1995
Robert Sambell
IT Engineer
b 16/07/1964
22/09/2007
Lucy Kiss
b 04/03/1975
Zachary Sambell
b 04/04/2000

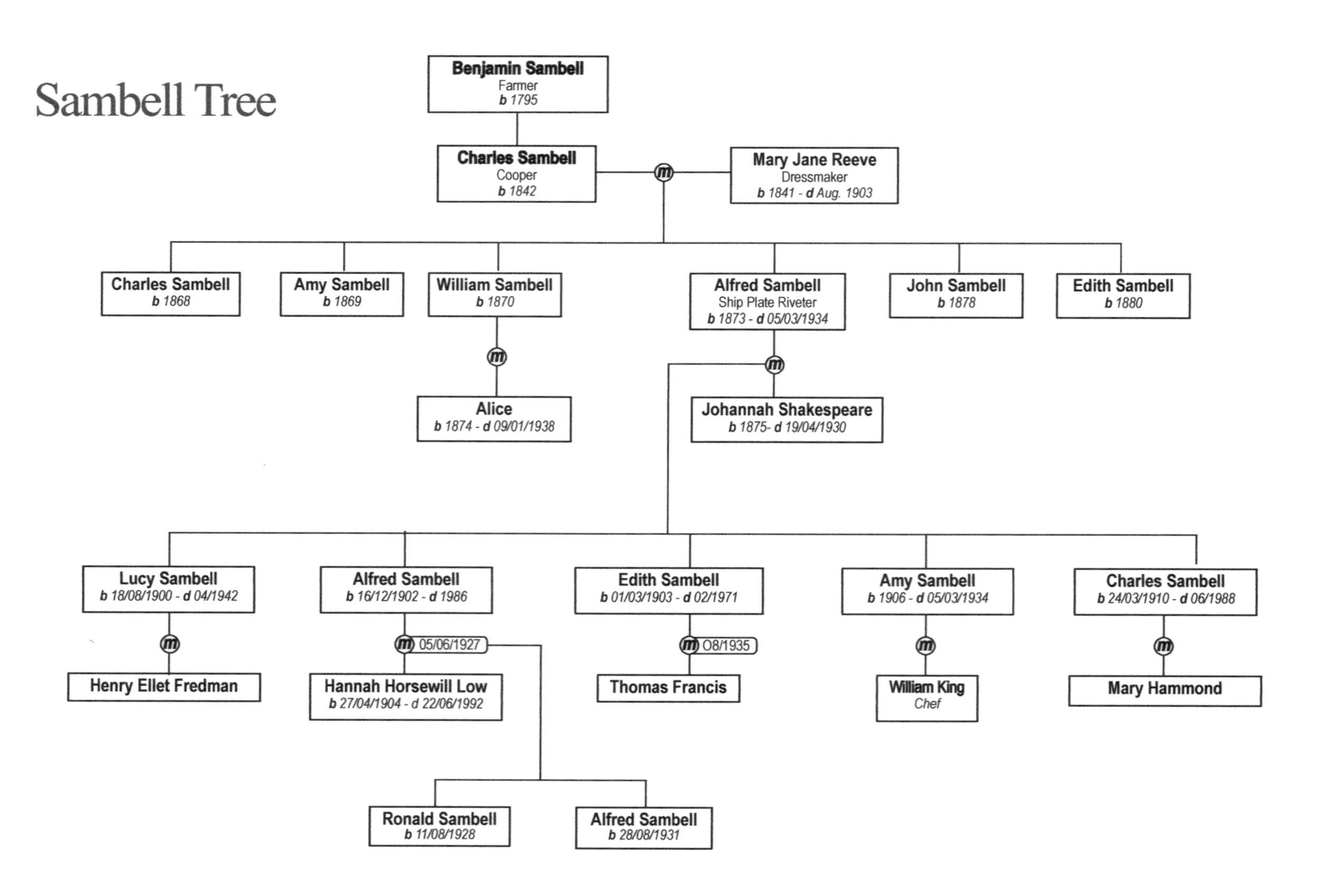
Sambell Tree
Benjamin Sambell
Farmer
b 1795
Charles Sambell
Cooper
b 1842
m
Mary Jane Reeve
Dressmaker
b 1841 - d Aug. 1903
Charles Sambell
b 1868
Amy Sambell
b 1869
William Sambell
b 1870
m
Alice
b 1874 - d 09/01/1938
Alfred Sambell
Ship Plate Riveter
b 1873 - d 05/03/1934
m
Johannah Shakespeare
b 1875- d 19/04/1930
John Sambell
b 1878
Edith Sambell
b 1880
Lucy Sambell
b 18/08/1900 - d 04/1942
m
Henry Ellet Fredman
Alfred Sambell
b 16/12/1902 - d 1986
m 05/06/1927
Hannah Horsewill Low
b 27/04/1904 - d 22/06/1992
Ronald Sambell
b 11/08/1928
Alfred Sambell
b 28/08/1931
Edith Sambell
b 01/03/1903 - d 02/1971
m O8/1935
Thomas Francis
Amy Sambell
b 1906 - d 05/03/1934
m
William King
Chef
Charles Sambell
b 24/03/1910 - d 06/1988
m
Mary Hammond

Horsewill/Dodsworth Tree

Two branches that link to Hannah Horsewill Sambell (my mother)

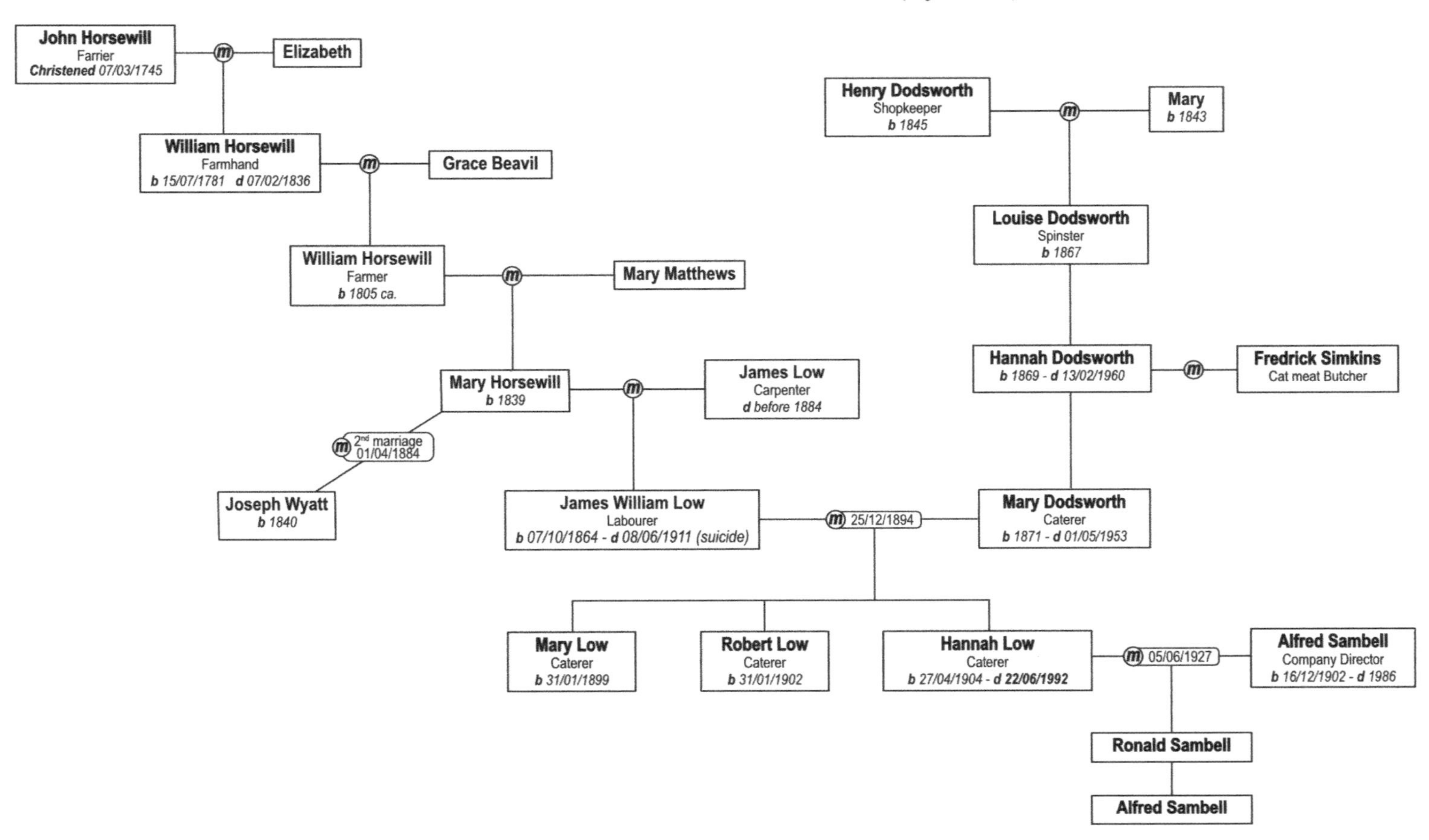

Ingram Content Group UK Ltd.
Milton Keynes UK
UKHW022054250423
420699UK00005B/151/J